PHILOSOPHY OF MEDICINE

Other interview books from Automatic Press ♦ $\frac{V}{I}$P

Formal Philosophy
edited by Vincent F. Hendricks & John Symons
November 2005

Masses of Formal Philosophy
edited by Vincent F. Hendricks & John Symons
October 2006

Political Questions: 5 Questions for Political Philosophers
edited by Morten Ebbe Juul Nielsen
December 2006

Philosophy of Technology: 5 Questions
edited by Jan-Kyrre Berg Olsen & Evan Selinger
February 2007

Game Theory: 5 Questions
edited by Vincent F. Hendricks & Pelle Guldborg Hansen
April 2007

Philosophy of Mathematics: 5 Questions
edited by Vincent F. Hendricks & Hannes Leitgeb
January 2008

Philosophy of Computing and Information: 5 Questions
edited by Luciano Floridi
Sepetmber 2008

Philosophy of the Social Sciences: 5 Questions
edited by Diego Ríos & Christoph Schmidt-Petri
September 2008

Epistemology: 5 Questions
edited by Vincent F. Hendricks & Duncan Pritchard
September 2008

Mind and Consciousness: 5 Questions
edited by Patrick Grim
January 2009

Philosophy of Science: 5 Questions
edited by Robert Rosenberger
November 2010

See all published and forthcoming books in the 5 Questions series at
www.vince-inc.com/automatic.html

PHILOSOPHY OF SCIENCE
5 QUESTIONS

edited by

Jan Kyrre Berg Olsen Friis
Peter Rossel
Michael Slot Norup

Automatic Press ♦ VIP

Automatic Press ♦ $\frac{V}{I}$P

Information on this title: www.vince-inc.com/automatic.html

© Automatic Press / VIP 2011

This publication is in copyright. Subject to statuary exception
and to the provisions of relevant collective licensing agreements,
no reproduction of any part may take place without
the written permission of the publisher.

First published 2011

Printed in the United States of America
and the United Kingdom

ISBN-10   87-92130-40-2   paperback
ISBN-13   978-87-92130-40-2   paperback

The publisher has no responsibilities for
the persistence or accuracy of URLs for external or
third party Internet Web sites referred to in this publication
and does not guarantee that any content on such
Web sites is, or will remain, accurate or appropriate.

Typeset in $\LaTeX 2_\varepsilon$
Cover design by Vincent F. Hendricks

Contents

Preface	iii
Acknowledgements	v
1 Vilhjálmur Árnason	1
2 Daniel Callahan	19
3 Arthur L. Caplan	29
4 Ruth Chadwick	31
5 Bill (KVM) Fulford	37
6 Henk ten Have	55
7 Bjørn Hofmann	79
8 Søren Holm	91
9 Ingvar Johansson	101
10 Niels Lynøe	111
11 Ruth Macklin	127
12 Lennart Nordenfelt	143
13 Onora O'Neill	163
14 Peter Rossel	171
15 Udo Schuklenk	189
About the Editor	205

Preface

Philosophy of Medicine: 5 Questions

◆

Although medicine and philosophy has been closely connected through most of their history philosophy of medicine as a distinct academic discipline did not emerge before the late 1960'es or early 1970'es. This new discipline had several sources. Within medicine a rapid and successful development involving artificial life support, organ transplantation, new genetic knowledge and new reproductive methods among other things raised a series of ethical and conceptual questions. At the same time changes in society gave way for a critique of traditional medical arrogance and paternalism, which emphasized the patient's autonomy and right to be informed and involved in the medical decisions. Further, many philosophers wished to move away from the sterile meta-ethics that had been dominant in ethics in the period before and confront real life problems of the kind that medicine could deliver.

From the beginning philosophy of medicine has engaged both philosophers and physicians in the analyses of ethical questions, both in the clinic and in relation to research, in the analysis of central concept in health care, in clarification of the rational basis for medical decision making and in illuminating basic questions in medical epistemology and methodology. Later development has broadened the scope and ethics of public health, global bioethics and international justice in relation to health has become important parts of the field within the last 10 years.

From the beginning philosophy of medicine has engaged both philosophers and physicians in the analyses of ethical questions, both in the clinic and in relation to research, in the analysis of central concept in health care, in clarification of the rational basis for medical decision making and in illuminating basic questions in medical epistemology and methodology. Later development has broadened the scope and ethics of public health, global bioethics and international justice in relation to health has become important parts of the field within the last 10 years.

The main focus of the majority of the contributors is on one or more of the many aspects of medical ethics or bioethics but there are also some who provide valuable knowledge of other parts of the broad landscape of medicine of philosophy e.g. the concept of disease. The volume is written in a language accessible to readers without training in the field and will work as a resource for students and researchers as well as for a broader audience interested in philosophy of medicine.

Jan Kyrre Berg Olsen Friis,
Peter Rossel and
Michael Slot Norup
Copenhagen / October 2011

Acknowledgements

We are particularly grateful to the contributors for devoting time to writing such erudite, enlightening and often thought-provoking interviews and grateful to the philosophical community in general for showing interest in this project. In addition, we would like to thank editor-in-chief Vincent F. Hendricks and associate editor Henrik A. Boensvang of Automatic Press ♦ $\frac{\vee}{\mid}$P.

Copenhagen / October 2011
Jan Kyrre Berg Olsen Friis,
Peter Rossel and
Michael Slott Norup
Editors

1
Vilhjálmur Árnason

Professor of philosophy

Department of Philosophy and Centre for Ethics, University of Iceland

---

1. Why were you initially drawn to Philosophy of Medicine?

My interest in the philosophy of medicine was awakened relatively late. I was from the beginning of my philosophical studies interested in ethical issues in a broad sense as they are addressed, for example, in existential and political thought. At the same time I was put off by mainstream moral philosophy which I found too much bogged down in unimportant technical details. It may be of relevance in this context that I was raised in a small fishing village, Neskaupstadur, on the east coast of Iceland. It was rather uncommon that young people would go for university education, let alone in such a 'useless' subject as philosophy. I have always, however, regarded philosophy as a most practical discipline, both as a mode of thinking in general and as a way to approach particular issues that are of concern to people. I was first introduced to applied ethics when I was an assistant in courses on animals ethics and business ethics in graduate school in the US, but I did not get acquainted with medical philosophy. In my doctoral dissertation, which was a critical appraisal of existential ethics in light of philosophical hermeneutics and critical theory, I argued for the need to place moral analysis in a larger social context without relating it to any particular social domain or practice.[1]

When I started teaching in the department of philosophy at the University of Iceland I was assigned to teach a course, A philosophical introduction to the sciences, in the Department of Nursing. This course was part of a general program of raising critical

---

[1] Árnason (1982).

consciousness among students about the nature and presuppositions of the sciences and to reflect critically upon their own field of study. The core subject matter of the course is a philosophical discussion about critical thinking, about the nature of scientific research, theory and method, the division of the sciences, the social effects of scientific activity, and the relationship between science and ethics. Additional subject matter depends on the needs and interests of the individual departments. In the case of nursing, the course should be oriented towards the ethics of nursing, medicine and health care. The specific objective of this course was to enable students to perceive, analyze and discuss ethical problems that arise in the nursing profession. I soon realized what a gold mine this field was for a philosopher who wanted to deal with matters of grave concern for ordinary people. In many ways I agreed with Stephen Toulmin's point about "how medicine saved the life of ethics" by providing the latter with "new preoccupation with situations, needs and interests" which require a careful analysis and practical reasoning.[2] I do not, however, share Toulmin's sharp contrast between 'the ethics of cases' as contrasted with 'the ethics of rules and principles' which I take to be a false dichotomy as I will touch upon below.

After teaching this course for a few years, I took my first sabbatical which I spent at Pacific Lutheran University in Tacoma, Washington, USA, in 1988. One of the teachers in the department was Paul Menzel who was working in medical ethics.[3] Paul took me to a lecture series which was held in the fall semester at the University of Washington, Seattle, called "The Ethicists: Life and Death on the Medical Frontier". Prof. Albert R. Jonsen was in charge of the series which featured eight of the most prominent people in the field of medical ethics. During my sabbatical I started writing a book in Icelandic about the issues in the ethics of medicine and health care.[4] The writing of the book took me five years and in the process I got well versed in the mainstream discourse about philosophy of medicine, especially biomedical ethics. In the spring of 1990, I attended the so-called "Extended European Bioethics Course" at the Joseph and Rose Kennedy Institute of Ethics at Georgetown University, Washington DC. The course

---

[2] Toulmin (1986), p. 266.
[3] Menzel (1990).
[4] The Icelandic title of the book could be translated as Ethics of Life and Death. Difficult Decisions in Health Care. It has been published in German: Árnason (2005), Dialog und Menschenwürde. Ethik im Gesundheitswesen.

was directed by Hans–Martin Sass, Robert M. Veatch and LeRoy Walters, and among other teachers were Tom Beauchamp, James Childress, Edmund Pellegrino and Ruth Faden.

2. What does your work reveal about Philosophy of Medicine that other academics, citizens, or economists typically fail to appreciate

I believe that the most distinctive feature about my reasoning in medical ethics has been the emphasis on the crucial and diverse role of dialogue, both in facilitating the professional-patient interaction, in securing informed consent, in fleshing out the idea of scientific citizenship, and facilitating just health care policy.

When I entered the scene of medical philosophy, the discussion about professional-patient interaction was characterized by a critique of traditional paternalism, most often providing a model of patient autonomy as an alternative vision. According to this model, the role of the professional was mainly to provide medical information upon which the patient could base her informed decision regarding treatment or research. This should preferably be done in a non-directive or even neutral way in order to free the patient from the values, goals and other controlling influences of the medical professionals who should limit their role to the medical and technical aspects of the situation. This was appropriately called the 'engineering model' in medicine and the label 'nurse technician' was also introduced.[5]

I argued that the patient autonomy model did not provide a desirable alternative to traditional paternalism. Despite their differences, the two models share in effect a major characteristic which has questionable consequences for the patient. Each in its own way, these models are monological in the sense that they emphasize either the professional communication of medical information or the patient's communication of their personal values and preferences. Neither model facilitates conversations or dialogue between patients and professionals. While the paternalistic model tends to ignore the patient's right to autonomous decision making, the patient autonomy model threatens to leave patients alone in their deliberations which can cause a feeling of abandonment, anxiety and a loss of trust which may undermine their decision making abilities.

These shortcomings cannot be properly amended by modeling

---

[5] Veatch (1981) and Smith (1981).

the interaction on a contract which protects the interests of both parties but is primarily guided by the preferences of the patient.[6] This contractual model is still based on an idea of rather strong autonomy of the patient which is appropriate in a more business like setting of customer-service provider or client-advisor than in the patient-professional interaction. Enclosing the partners within their private sphere of value preferences whose integrity rests on not having to compromise them in any way, the contractual model bypasses a major asset of a dialogical model: When people really engage in a conversation, they open up to each other and to themselves in such a way that their preferences are tested and possibly reevaluated in the process.[7]

In order to overcome the shortcomings of these models of interaction, I attempted to develop a communicative model which aims at building up mutual trust and responsibility of both partners.[8] Both cognitive and emotional factors can disrupt autonomous decision making and a true dialogue where people meet in a joint task can serve as a midwife of good decision making.[9] Good communication in medical practice, therefore, has two main objectives: information or freedom from ignorance, and emotional support or freedom from fear and anxiety. Such a dialogue, if authentically conducted and aimed at mutual understanding, also breaks up institutional routine because it takes time and is not subject to the demands of efficiency and control. It is also the best way to build trust which to many patients is more important than the exercise of self-determination. If our notions of patient decision-making are to be useful they must relate in a realistic way to the situation and experiences of patient. As the psychiatrist Jay Katz argued in an excellent book, professionals know more about the treatment or study, patients know more about themselves.[10]

I have found the role of the dialogue particularly important in the discussion of informed consent. The idea of informed consent implies that patients are informed, that they understand the information and that they make a voluntary decision. However, often patients are distressed, the information complex, and the hospital environment and the busy routine of health care professional not conducive to the kind of deliberation that needs to precede

---

[6] Ibid.
[7] Árnason (2005), esp. ch. 2.6.
[8] Árnason (1994).
[9] Árnason (2000).
[10] Katz (1984).

an informed, voluntary decision. As a consequence, the practice of informed consent tends to be reduced to the act of signing an informed consent form. Such a formal act meets some administrative requirements but need not entail any of the elements of genuine informed consent. The experience of this practice has understandably bred cynicism towards informed consent among both patients and practitioners. If the standards of genuine informed consent cannot be met while the practice of medicine and research is largely justified by it, the result will be "systematic hypocrisy".[11]

A sensible response to this dilemma is to emphasize that obtaining informed consent is not an event but rather a communicative process between practitioners and patients.[12] Engaging in communication implies listening no less than talking to patients. The best way to find the adequate disclosure of information for a particular patient is to have a dialogical exchange of questions and answers. Only in this way, can professionals know what information patients care for and need to have and what they do not. Also, a conversation between patient and professionals will reveal better than other available means whether the patient has understood the information or not. Communication does not only convey information, it also provides support and thus meets the needs for counseling and comfort many patients have. Such a communicative approach can thus be sensitive to cultural differences and be understanding of those who do not share common assumptions. To be sure, such conversations take time but they also build mutual trust which can remove many obstacles from the practice of informed consent. If the practical exigencies of medicine make such communicative effort unrealistic, then it amounts to admitting that there cannot be good medical practice.[13]

Obviously, it is more difficult to facilitate dialogical conditions for informing participants in non-clinical research which requires large amount of participants. An interesting example which has posed challenging questions about the possibility of obtaining informed consent is population genetic database research which I have been concerned with for some time.[14] My reflection on these issues has mainly been inspired by the Icelandic experience of hav-

---

[11] Manson and O'Neill (2007), p. 25.
[12] Wear (1998), pp. 178–179.
[13] Árnason, Li and Cong (forthcoming).
[14] Cf. Árnason (2004); Árnason and Árnason (2004), Kristinsson and Árnason (2007).

ing one of the world's largest genetics research company which has built up population databases for its research.

The prevailing positions on this matter can be divided into three main categories. First, there are those who argue for the need to inform patients specifically about each particular research project planned in a database. Such specific consent implies that participants will be informed prior to donating their data for research about its objectives, methods, risks, benefits and other traditional ingredients of informed consent. This means that any research with new questions requires re-contact with the participants. This severely restricts the flexibility of database research without contributing much to participants' understanding who find this policy annoying and are willing to give a more open consent.

The second position emphasizes that population databases are resources for genetic research and it is impossible to describe in detail the research that will be performed on the data at the time of collection. If we are to use this resource efficiently, specific consent for database research must be rejected in favor of an open consent. By an open consent is meant here that participants agree that their data will be used for any future scientific research permitted by regulatory institutions. However, such an open consent does not provide participants with the information necessary for them to make a meaningful choice, i.e. act in a voluntary way on a basic understanding of the matter. It transfers the reflection on population research from the participants to regulatory institutions. Thus motivations for scientific literacy and awareness of the public would be reduced and important benefits related to human agency are ignored.

I have found it important to avoid these pitfalls of specific and open consent and have taken part in attempts to carve out alternatives that are intended to strike a balance between the researchers' need for flexibility and the ethical demand for protection of participants' interests.[15] The main thrust of these proposals, which have different emphasis, is that participants should be asked to authorize the use of their data for described health care research. They would be informed about the conditions for use of the data, such as how the research will be regulated, how they will be connected to other data, who will have access to the information, how privacy will be secured, and that they will only be used for

---

[15] Greely (1999); Caulfield, Upshur, Daar (2003); Árnason (2004); Kaye (2004).

described health care purposes. Participants would be told that they and/or their proxies will be regularly informed about the research practice and that they can at any time withdraw from particular research projects. In this way, the emphasis on a one time initial consent is rejected in favor of a dynamic dialogical process between researchers and participants.

In this way, the idea of the dialogue is translated into a communicative practice in population database research. In addition to being the most fitting way for such research, it also protects better than other models interests associated with moral agency while securing the moral purposes of informed consent. It is my contention that these interests have been neglected in the discussion about population genetics, biobanks and databases. Recently, my research has been focusing on the idea of democratic biopolitics, showing how the interest in agency has been thrust aside in favor of protection and benefit reaping.[16]

The ethical regulation of biotechnology and research on humans emphasizes above all the protection of people from the possible misuse of information and technology. Some of the major moral objectives in research ethics are protection of privacy, protection against risks (participants' welfare) and protection of vulnerable research subjects (a major requirement of justice). In all these cases, measures are to be taken that safeguard research participants and citizens in general. This is an important view but it is limited. The other prevailing view places the emphasis on the benefits that can be reaped from biotechnology and genetic research. These benefits can be either health-related, such as drug development, more effective predictive and preventive medicine, or benefits unrelated to health, such as increased employment opportunities for young scientists and other social and economic advantages that may flow from having thriving research companies. This medical and social utility position has been prevailing in political and economic discourse about biotechnology.

This argument from collective interests is often used to accuse the protective position of emphasizing individual rights at the cost of social goods. The emphasis should be on the duties of participants to contribute to progress in medicine and science no less than on having their privacy rights protected. However, there are important public interests at stake as well in maintaining the ethos of voluntary consent to participation in database research and ne-

---

[16] Árnason and Hjörleifsson (2007b); Árnason (2008), Árnason (forthcoming).

glecting it may weaken a democratic society in the long run. Thus motivation for scientific literacy and awareness among the public would be reduced which is not in the public interest. The benefit view as I have described it thus ignores important benefits related to human agency. From this perspective, the sharp distinction between individual and collective interests is misleading. Providing options for participants' deliberation and preserving other conditions for human agency and reflection are not mere private interests. However, these objectives are not best served by obtaining specific informed consent from otherwise passive participants.

In the context of my discussion of the relationship of these views to democracy and citizenship, their common shortcomings and limits have become more conspicuous than their differences. Both positions disclose an important underlying and hidden presumption concerning the scientific citizenry that is being created. By placing the main emphasis either on protecting the participants' private domain from illegitimate interference or on providing them with material benefits, these positions regard people in a passive role. They do not provide reasons for implementing policies that facilitate deliberation in the public sphere. In this way they are part of a research culture which contributes to scientific illiteracy and disregards the active elements of human agency which are crucial for democracy. I have argued for the need, therefore, to look for ways to increase public awareness of population research and strengthen the conditions for their decisions and responsibility for participation in research.

I have formulated this alternative in my argumentation for the authorization model of consent for population databases. This model provides conditions for an active opt-out clause which is likely to create more informed and critically aware citizens and is also conducive to informed trust. This position enables active scientific citizenship because it emphasizes the creation of conditions or opportunities for citizens to reflect on their participation in scientific research. Contrary to the protective policy of specific informed consent, these conditions for participants' deliberation do not come at the cost of a flexible biobank research. There is no requirement of a continuous re-consent in order to meet formal procedures, but a dynamic interchange which has the primary aim of keeping participants informed and aware. Hence, such scientific citizenship needs not thwart the possibilities of reaping the benefits of biobank research but it avoids reducing participants to being merely passive part of a resource. The objective is to create more informed or educated citizens who do not have to rely exclusively on expert knowledge but can use it in their deliberations about research participation. This, of course, is not something that can be easily realized but it is an important vision to guide our attempts in shaping citizens' awareness in society where biological research and biotechnology play an increasing role.

Finally, I have put the idea of dialogical deliberation to use in my analysis of just health care and attempts to work out a more just health care policy. In dealing with macro-allocation of health care goods and the making of general health policy, it is necessary to form a social agreement about general principles. Even though they are not substantiated in health care policy, reports about prioritization in health care from the Nordic countries, for example, harbor important indicators of the citizens' views about what must guide us in this task. In the task of setting just limits to health care, our alleged adherence to justice and solidarity is put to the test because we can no longer escape the task of finding out which inequalities are justifiable and which are not. I have argued that in order to shape a mode of thought about the health care system which can fairly limit our access to health care goods without violating solidarity, we can look in the direction of Rawls' theory of justice as fairness.[17]

For this purpose, we must not read Rawls' theory too narrowly as a contract theory and regard the principle of justice as fair simply because they are chosen under a veil of ignorance. Contrary to many critics, the rootedness of the theory of justice as fairness in democratic culture shows that even the liberal contractual system of solidarity is founded upon a pre-contractual consensus on fundamental values. On this reading one can see the hypothetical contractors under a veil of ignorance choose to design a solidaristic system of health care not simply because they are rational egoists but also because they realize the dependence of individuals upon a system of social relationships and values.

The function of the veil of ignorance is not only to make the contractors ignorant of their own position but also to make them more knowledgeable about the human condition in general and about various individual situations. As a theoretical exercise it inspires us to imagine ourselves in the situation of the worst off and thus it motivates our vision of interdependence and reciprocity

---

[17] Árnason (2008b).

in human relations.[18] In this vein of thought, Rawlsian justice can thus also contribute to "reflexive solidarity" which "implies continuous reappraisal of the way that institutions and services affect the people involved in caring practices".[19]

Surely, this is a procedural notion of justice, specifying the conditions necessary for a fair distribution of goods in society. But it can also be seen as a critical idea, providing a perspective from where every real agreement or consensus in society can be normatively assessed. Such a consensus requires societal dialogue which ensures that people take responsibility for the policy made, and identify with it. For this discursive task, Rawls' theory may not provide sufficient guidance because it ignores the ongoing democratic dialogue where the principles of justice themselves must be open for revision.[20] Nevertheless, we can learn from it that in fleshing out the idea of a social consensus on health care we are bound to move from individual rights-based attitude to general rules of fairness which set limits to health care that everyone can in principle accept. Therefore, ideas put forth in official reports need to be tested in public deliberation in the spirit of communicative ethics and democracy. A political legitimation in the spirit of deliberative democracy can only be reached by a preceding critical discussion in the public sphere the outcome of which is translated into political will formation. This critical idea of freedom in public deliberation needs to be taken more into account in the exercise of deliberative democracy if it is to contribute to overcoming the limits of the protecting and the benefit positions in bioethical discourse and create conditions for more informed and engaged citizens.

3. What, if any, practical and/or social-political obligations follow from studying medicine from a philosophical point of view?

It is implied in the previous discussion about health policy that the role of the moral philosopher is to engage in the social debate about health care. In order to be of use in such debate, it is important to find a proper balance between what could be called a stance of participation on the one hand and a position of distanciation on the other hand. By participation in this context, I

---

[18] Okin (1989).
[19] Houtepen and ter Meulen (2000), 373.
[20] McCarthy (1994).

mean that it is necessary for the participant in a constructive debate about medicine and health policy to be knowledgable about the subject matter and have a feeling for the practice and policy being discussed. On the other hand, it is just as necessary for the philosopher to keep a critical distance to the practice of medicine. As I see it, this relates to the question about the nature and role of bioethics. Obviously, attentiveness to experience, practice and context is crucial to a successful moral analysis and I welcome attempts to reconsider bioethics in order to strengthen these factors as is commonly done in the movement of so-called empirical bioethics. However, I have been critical of views about the role of moral philosophy in interdisciplinary research which emphasizes the primacy of accepted norms over reasoned principles.

I believe that the practical obligation that follows from studying philosophy of medicine or medical ethics is to practice critical moral thinking for example in sensible bioethical discussion.[21] We exercise the power of speech in such discussion which should not have any apriori theoretical restrictions other than what is required by its inherent rationality and good scientific practice. Free moral thinking must not be told to respect the existing norms because such thinking inevitably implies that the reasons for the normative claims made shall stand to scrutiny. Just as empirical science must critically examine truth claims, so ethics needs to critically scrutinize claims to rightness which are embodied in law and local standards, in actual ethical views or an established social consensus. It is necessary to understand the normativity which resides in the social context but this is never sufficient to conclude moral reasoning. Surely, the norms for reflection are found in actual practice but the reasoning about their validity should not be limited by the normative implications of an established practice. The question, therefore, is not only what is in fact accepted but also why it is accepted and whether it is worthy of recognition.[22] It is elementary for moral analysis that the fact that people accept something or that it has been enacted in law or declarations does not imply that there are good reasons for accepting it. This is the main reason why moral analysis must not be too close to the practice itself, however contextually sensitive and participating it wants to be, because reflective distanciation is the precondition for a fruitful moral investigation.

---

[21] Árnason (2005b).
[22] Habermas (1990), p. 61.

Moreover, in sensible moral thought, moral principles are not rigidly applied; they inform moral reasoning, help identify relevant features of situations and enlighten decisions. From this perspective, "principles are not instructions to avoid examining particulars, but rather are instructions about what to look for".[23] An obvious critique of this position from a contextualist point of view is that the interpretative framework itself is skewed towards dominance of principles which is bound to distort the local contexts which are foreign to principled thinking and thus instruct us to look in the wrong directions. I find two things of main importance to respond to this point. One is that the values which the principles receive their significance from, must always be interpreted in light of the cultural context under scrutiny. The other is that even though the values tend to be culturally different they relate to underlying human interests which are generalizable in the sense that they are shared by human beings, even though they take on different cultural manifestations[24].

4. What do you see as the most interesting criticism against your own position in philosophy of medicine?

The most interesting criticism of my position in the philosophy of medicine, or the one that I have found most instructive and useful for the development of my thought even though it has not been aimed directly at my writings, has to do with the danger of bioethics inadvertently assuming a legitimating and an ideological role in the discussion about new biotechnology or policies. Bioethics can take on a legitimating role by focusing too narrowly on particular ethical questions at the neglect of the larger social implications.[25] Bioethical analyses tend to focus on questions concerning a particular set of issues relating to basic human interest, such as privacy and consent, risk of harm or discrimination. If there are good reasons to believe that these interests can be protected, a particular bioethical technology could legitimately be introduced. For example, the introduction of genetic testing could be discussed primarily in terms of the main principles of autonomy, non-maleficence, beneficence and justice, evaluating whether the practice would duly meet the requirements for informed consent of patients undergoing the tests, whether the test would put the

---

[23] Kymlicka (2002), p. 404.
[24] Árnason (2006).
[25] Árnason and Hjörleifsson (2007a).

person at considerable risk, e.g. relating to knowledge of nontreatable disease, whether privacy of information would be protected so that the patient would not be in danger of being discriminated against on the basis of his genetic susceptibility for certain diseases. The question concerning the effects of introducing pervasive genetic testing upon health care services and the practice of medicine would not, however, be taken into account. Such a narrow ethical discourse is ideological in the sense that it implicitly covers up important moral aspects of the effects of biotechnology while claiming to analyze its main ethical implications. This is a major reason why I have paid an increasing attention to social and political issues and become more convinced that bioethics must not be distinguished sharply from biopolitics.[26]

This criticism of mainstream bioethics is, however, often taken too far by implying that the principled approach in bioethics is inherently individualistic. The relationship between attending to socio-political matters and the prevailing trend in mainstream bioethics which is usually referred to as 'ethics of principles' is far from obvious. It partly depends on how the principles are understood and how they are used in bioethical discussion. The issue of autonomy, for example, is a complex philosophical matter and its identification with individual choice based on informed deliberation is only one expression of it and should not be taken as a universal model, even though it has been prominent in mainstream biomedical discussion. It could even be argued that this particular mode of respecting autonomy can in some cases undermine agency which is the underlying interest of the principle. The 'problem' with the four Georgetown principles, for example, is not which principles are put forth but rather that they are often simplistically applied and identified with a certain individualistic interpretation of the underlying values.

I see no inherent tension between emphasizing the value of autonomy or other dominating ideas of 'mainstream bioethical discourse' and the heuristic reminder to pay attention to larger socioethical issues. To the contrary, it is one of the preconditions of a rich notion of autonomy not to reduce it to choices which are abstracted from the sociocultural context framing these choices. It is only when autonomy is interpreted in an atomistic way that it leads to a neglect of wider biopolitical and cultural issues. This shows how important it is to broaden and deepen the philosophical

---

[26] Árnason (2008a).

reflection which forms the basis of medical ethics.

5. With respect to present and future inquiry, how can the most important problems concerning Philosophy of Medicine be identified and explored?

I have already mentioned the importance of placing biomedical ethics more in the context of social and political issues. But I would also like to see developments in philosophy of medicine that would influence biomedical thinking and bioethical practice. It is often said that with the increasing objectification of the patient that accompanies the ever more sophisticated specialization of medical doctors in parts of the body, medical ethics has a major function to salvage the subjective aspect, the patient as a person. While this is an important message, it also entails a cleavage between the scientific task of understanding the body and acquiring technical mastery over it on the one hand and the moral task of protecting human values on the other hand. This separation of the medical and the moral covers up the fact that medicine is inherently an ethical discipline because the subject matter is the human being in its multifaceted relationship with himself, other persons and the environment at large. And it is crucial not to limit this relationship to the subjective aspect of the human person but extend it to the physical as well, or rather to understand the physical in its complex interaction with the existential and the cultural dimensions of the human being. Perhaps the most important task in laying the foundation of a richer medical anthropology is a rethinking of the human body. Such a rethinking would not only have implications for a scientific understanding of the body but also for clinical practice and on the general way in which the problems of the patients are framed, perceived and dealt with. Furthermore, a more humanistic or holistic medicine would have to integrate the ethical dimensions at every level of medical practice. The ethical would no longer be an epiphenomenal aspect of medicine but integral to it. Such a reconsideration would obviously have major implications for medical education, for example by rejecting the departmentalization of the ethical and medical and putting an end to the "quarantine of philosophy" in medical education.[27]

---

[27] Cf. William E. Stempsey. "The Quarantine of Philosophy in Medical Education: Why Teaching the Humanities May Not Produce Humane Physicians." Medicine, Health Care and Philosophy 2 (1999): 3-9.

This task of revising the subject matter of medicine would require an intensive and ongoing dialogue between physicians, medical philosophers and other thinkers and scientists of medicine. It is quite possible, however, that if this vision of a new medical anthropology was to be realized it would do away with medical ethics as we know it, i.e. as a discipline separate from medicine. But this is a common feature of most ambitious aspirations. If all the dreams of a just society and prosperous life were to come true, there would be no need for traditional ethical thinking. Perhaps it is one indication of whether we are heading in the right direction that the utopian reference implicitly appealed to implies the dissolution of our current modes of thought and practices.

References

Árnason, Vilhjálmur (1982). The Context of Morality and the Question of Ethics. From Naive Existentialism to Suspicious Hermeneutics. Purdue University, University Microfilms, Ann Arbor.

Árnason, Vilhjálmur (1994) "Towards Authentic Conversations. Authenticity in the Patient-Professional Relationship." Theoretical Medicine 15, 227–242.

Árnason, Vilhjálmur (2000). "Gadamerian Dialogue in the Patient-Professional Interaction". Medicine, Health Care and Philosophy 3, 17–23.

Árnason, Vilhjálmur (2004). "Coding and consent: Moral challenges of the database project in Iceland", Bioethics 18: 27–49.

Árnason, Vilhjálmur and Gardar Árnason (2004). "Informed, Democratic Consent? The Case of the Icelandic Database." Trames 8 (2004), 164–177.

Árnason, Vilhjálmur (2005a). Dialog und Menschenwürde. Ethik im Gesundheitswesen. German translation Lúðvík E. Gústafsson. Münster: Lit-Verlag.

Árnason, Vilhjálmur (2005b). "Sensible Discussion in Bioethics. Reflections on Interdisciplinary Research", Cambridge Quarterly of Health Care Ethics14, 322–328.

Árnason, Vilhjálmur (2006). "The Global and the Local. Fruitful Tension in Medical Ethics". Ethik in der Medizin 18, 385–389.

Árnason, Vilhjálmur and Stefán Hjörleifsson (2007a). "Geneticization and bioethics: advancing debate and research". Medicine, Health Care and Philosophy 10, 417–431.

Árnason, Vilhjálmur and Stefán Hjörleifsson (2007b). "Population Databanks and Democracy in Light of the Icelandic Experience". Genetic Democracy. Philosophical Perspectives. Veikko Launis and Juha Räikkä (eds.). Springer Verlag, 93–104.

Árnason, Vilhjálmur (2008a). "Biopolitics in a Democratic Society." Bioethics, Politics and Business. TemaNord 570, 15–26.

Árnason, Vilhjálmur (2008b). "Justice or Solidarity? Thinking about Nordic Prioritization in Light of Rawls". In: Cutting Through the Surface: Philosophical Approaches to Bioethics, Søren Holm, Peter Herissone-Kelly and Tuija Takala (eds). Rodopi.

Árnason, Vilhjálmur, Hongwen Li and Yali Cong (forthcoming). "Informed Consent". In: Health Care Ethics in an Era of Globalisation, Ruth Chadwick, Henk ten Have, Eric Meslin (eds.) Sage Handbook Series.

Árnason, Vilhjálmur (forthcoming). "Scientific citizenship, benefit, and protection in population based research." In: Ethics of research biobanking. Jan Helge Solbakk, Björn Hoffman and Sören Holm (eds.). Springer Verlag.

Caulfield, Tim, Ross Upshur, Abdallah Daar (2003): "DNA Databanks and Consent: A Suggested Policy Option Involving an Authorization Model", BMC Medical Ethics 4.

Greely, Henry (1999). "Breaking the Stalemate: A Prospective Regulatory Framework for Unforeseen Research Uses of Human Tissue Samples and Health Information." Wake Forest Law Review 34: 737–766.

Kaye, Jane (2004). "Broad Consent — the only option for population genetic databases". In: Blood and Data: Ethical, Legal and Social Aspects of Human Genetic Databases, Gardar Árnason, Salvör Nordal, Vilhjálmur Árnason, (eds.) University of Iceland Press), 103–109.

Habermas, Jürgen Moral Consciousness and Communicative Action, trans. C. Lenhardt and S.W. Nicholsen (Polity Press 1990), p. 61.

Houtepen, Rob and Ruud van ter Meulen (eds.) (2000). Solidarity in Health Care, Health Care Analysis 8, Special Issue.

Katz, Jay (1984). The Silent World of Doctor and Patient. The Free Press.

Kristinsson, Sigurður and Vilhjálmur Árnason (2007). "Informed consent and human genetic database research". In: The Ethics and Goverance of Human Genetic Databases. European Perspectives. Matti Häyry, Ruth Chadwick, Vilhjálmur Árnason, Gardar Árnason (eds.) Cambridge University Press 2007), pp. 199–216.

Manson, Neil C. and Onora O'Neill (2007). Rethinking Informed Consent in Bioethics. Cambridge University Press.

Kymlicka, Will (2002). Contemporary Political Philosophy. An Introduction. Oxford University Press.

McCarthy, Thomas "Kantian Constructivism and Reconstructivism: Rawls and Habermas in Dialogue", Ethics 105 (1994), pp. 44–63.

Menzel, Paul (1990) Strong Medicine. Oxford University Press.

Okin, Susan Moller (1989). "Reason and Feeling in Thinking about Justice", Ethics 99 (2), 229–249.

Smith, Sheri (1981). "Three Models of the Nurse-Patient Relationship". In: Mappes, Thomas A. and Zembaty, Jane S. (eds.), Biomedical Ethics (McGraw-Hill), 120–126.

Stempsey, William E. (1999). "The Quarantine of Philosophy in Medical Education: Why Teaching the Humanities May Not Produce Humane Physicians." Medicine, Health Care and Philosophy 2, 3–9.

Toulmin, Stephen. (1986). "How medicine saved the life of ethics". In: J.P. DeMarco and R.M. Fox eds. New Directions in Ethics. The Challenge of Applied Ethics. Routledge and Kegan Paul, 265–281.

Veatch, Robert M. (1981). "Models for Ethical Medicine in a Revolutionary Age". In: Thomas A. Mappes,. and Jane S. Zembaty (eds.), Biomedical Ethics (McGraw-Hill), 56–59.

Wear, Stephen (1998). Informed Consent. Patient Autonomy and Clinician Beneficence within Health Care, 2nd. ed. Georgetown University Press.

1. Vilhjálmur Árnason

2

Daniel Callahan

Senior Researcher and President Emeritus

The Hastings Center

---

1. Why were you initially drawn to Philosophy of Medicine?

Let me begin by defining what I take to be the "philosophy of medicine," which I believe has no clearly fixed meaning. By the term I mean the animating rationale, goals, and ideals of medicine as a practice and a social institution. I take medicine to be a humanistic discipline that makes use of science to pursue its ends. In saying that I mean to reject a common belief that medicine is a scientific discipline with some secondary humanistic values and aims. At its core, medicine engages in the pursuit of health, that is, the well working of the body and the mind in order that the various goals of human life, individually and collectively, can be pursued. Understood in that sense medicine pursues the foundational goal of an important feature of the human good, that of bodily and psychological functioning sufficient enough to allow the pursuit of other human goods.

I was drawn to the philosophy of medicine because, over the years, my interest in some clinical and policy issues of medicine and health care (which encompasses but is not limited to medicine) inexorably pushed me in that direction. Let me offer three examples that led me on.

Death. The first was while doing work on end-life-care some 40 years ago. The emphasis in those early years of discussion was on the rights of patient when critically or terminally ill. The general response to that issue was that patients have to right to make their own decisions about terminating treatment; physician paternalism was rejected. But I felt that discussion, while important, was thin. Some more fundamental questions needed answering. How was death to be understood in the culture and traditions of medicine?

What was, so to speak, medicine's stance toward death? It seemed to me that modern science- and technology-driven had come to treat death as the ultimate enemy.

A major complaint by the 1960s and 1970s was that the care of the dying had come to be dominated by "the technological imperative: that it was the duty of the physician to use all available technology to keep patients alive and to do so regardless of the quality of their life. In the United States, on a parallel track, was our National Institutes of Health, which by the 1960s gave the highest research priority to the most widespread lethal diseases, notably cancer, heart disease, and strokes. A much lower priority was given to medical conditions that could generate misery but did not kill: such as arthritis, severe mental health problems, and osteoporosis.

In short, from both a clinical and research perspective, death had become the main villain. But it had become so in great part because, by the end of the 19th century, it became clear that medicine could do something about death. Fatalism should be replaced by scientific hope in the struggle against death and the idea of finitude replaced by utopian possibilities of changing the biological boundaries on life's possibilities. At the least I found that view semi-delusional, "semi" because much can be done to change human biology, but "delusional" because we will remain finite, of which death is now and always be, a part of human life.

Medicine, I concluded, must not take death to be its ultimate enemy. It can be an evil, but not always, and human society is not fundamentally harmed by the fact that people die, especially if they do so in old age and not prematurely. But contemporary medicine is schizoid: the clinician knows that patients die, and patients know that also, and that reality must be accepted–and yet that reality is not accepted by the research enterprise, which has declared all lethal diseases to be unacceptable and requiring elimination. I believe it is a human good to fight death, but by no means a moral obligation.

Progress. My interest in the place of death in medical philosophy (at least in some defacto, not always articulated, sense) led me to ask some other fundamental questions. Medical progress and technological innovation are core values of contemporary science-driven medicine and health care. But what kind of progress should be pursued and, for that matter, what should be counted as progress? Remarkably enough there is little literature in bioethics or philosophy of medicine on progress. It seems to be assumed that it is a self-evident value, needing neither justification or elucidation. A solid analysis of the concept of progress would, at the least, need criteria to determine what counts as valuable, dangerous, or trivial progress. If progress means no more than, as an outcome of medical research, that A is slightly more beneficial to patients than B, then there might be no problem. But if B is much more expensive than A, and thus a problem for health care financing, it might not be deemed progress at all from the perspective of public welfare even if beneficial for individual patients. Progress can be thought of in terms of gains in health for individuals, or in terms of gains in population health.

Health. There is a large and interesting literature on the concept of health, but it has rarely been used to help develop health policy. This may be partly because a consideration of health as a human value and thus of foundational importance for health care policy runs into many puzzles and paradoxes. People in good health can be unhappy and dissatisfied with their life, and many people in poor health can cope well with it and be satisfied with their life. The connection between health and a sense of well-being is hardly clear and often unpredictable for individuals dealing with illness. What people bring to illness can be as important as what illness brings to people. Trying to develop policies with these variables in mind is at best a difficult venture. Most typically a person is thought to be unhealthy if he has a disease or medical disorder that deviates from what Norman Daniels has called a "species typical" standard. But that determination tells us little about how the designated sick person will respond to his malady.

There is another and more recent medical phenomenon, that of the medicalization of larger swathes of human distress and discomfort, previously thought to be just life. They include male pattern baldness, sports medicine and medical repair of bad knees in aging athletes (professional or recreational), Viagra for waning sexual prowess, and contraception to avoid pregnancy (neither a disease nor a malady, just undesirable on occasion), and a growing tendency to treat species typical physiological decline in later life as if is a treatable pathology. In other words health is an expanding concept, a function of changed expectations, culturally induced shifts in the meaning of the related concepts of illness and disease, and successful medical interventions for previously non-medical conditions.

Implications for health policy. My own work in recent years has focused on attempting to bring our understanding of death, progress, and health into health policy discourse and formation. The impetus for doing so is the difficulty all health care systems in developed countries now have in dealing with the constant and steady rise in health care costs. The most common way of dealing with them is by the application of organizational and managerial strategies, on the one hand, or ideological strategies, on the other (privatization versus government dominance)—or some combination of both.

Those efforts are failing (though more in the US than in Europe), and there is no reason to think, after decades of efforts to control costs, that they will ever be successful.

The real difficulty is that the values contemporary medicine brings to health care (with full public support) are themselves the problem. How can a society control costs when it believes that death is the greatest enemy of life, forever to be fought against; when it believes that medical progress must always go forward; and when it believes that health is an elastic concept, open to endless expansion based on technological innovation and cultural redefinition? It can not do so under those circumstances. Death can be pushed back and forestalled, but not eliminated. The cost of pursuing endless progress has become prohibitive. Medical research almost invariably raises, not lowers, health care costs.

As progress improves health the standard of what counts as good health rises with it. The outcome of that phenomenon is that the healthier we become, more is spent on health care, not less. It is the dog chasing its tail. From an economic perspective that is a perverse outcome. Striving for infinite goals–with no end in sight that would count as success—with finite resources is a recipe for economic overload. Medical progress can be likened to the exploration of outer space: no matter how far we go there is still further that we can go. With space exploration, however, fiscal limits are recognized, and we settle for space stations and unmanned reconnaissance craft. We don't consider that limitation a tragedy, but we act as if it is a tragedy that people continue to die of disease. Sometimes it is and sometimes not.

In sum, I do not believe it possible to have sustainable and affordable health care systems without basing them on a careful incorporation of some fundamental issues in the philosophy of medicine.

2. What does your work reveal about Philosophy of Medicine that other academics, citizens, or economists typically fail to appreciate?

I think it fair to say that, over the years, I have been a much cited author in bioethics, and on a wide range issues, from abortion at the beginning of life to the care of the dying at the end. Yet I have been disappointed that my writing on death and progress seem to have had little impact. I am not sure just why that is, particularly since some of my books on those topics were well reviewed and even sold well. But they did not make their way far into the bioethical literature.

My own guess about that omission is two-fold. First, the philosophy of medicine as a field has not been much drawn upon to deal with policy issues; it seems to be treated as separate and independent, not oriented to policy questions. Second, contemporary bioethicists and moral philosophers are hesitant to plumb the depths of philosophy of medicine in the service of policy formation–that would make the policy task harder and not easier.It is easier and less controversial to settle for procedural solutions to the ethical dilemmas of policy than to draw upon susbstantive reflections on the philosophy of medicine. There seems, for instance, to be little hesitation to deal with end-of-life issues if they are limited to the rights of patients to make their own decisions. But it is hard to find anyone who has understood that patients, and the cultures of which they are a part, need to ponder just how to think about death itself. To say I should have a choice about my dying makes considerable sense. But how can I determine what would count as a good or bad choice, which requires that I have some grasp of the place and meaning of death in our lives.

As another example I have come to think that a just health care system can not be sustained without some consensus on what kind of a system, with what philosophical principles, will make justice even feasible. Two questions of justice must be answered together. What resources, with what goals in sight, should be devoted to health care? What would count as a fair and equitable distribution of those resources? Moral philosophers and political scientists have lavished considerable attention on the second question, but in minimizing attention to the first one have succeeded too often in creating an abstract and often sterile academic exercise with little obvious impact on policy. I think it impossible to meaningfully deal with justice in health care without coming to grips with the de facto research warfare against death and the embrace of infinite progress and technological innovation that fuels that warfare. Together they are the main driver of rising health care costs. A health care system with those values can not be fashioned into a just system, and that is exactly what is happening with the systems of developed countries, but most notably in the United States, where they are unexamined premises of the health enterprise.

3. What, if any, practical and/or social-political obligations follow from studying medicine from a philosophical point of view?

I find this a difficult question to answer, if only because I am not sure what a "philosophical point of view" is. Does it mean adopting the methodologies of philosophy, or following some particular school of philosophy? Or some particular style of reasoning and analysis? Since I am not sure about the answers to that question, let me just say a few things about the role of philosophy in medicine and health care.

I received my Ph.D. from Harvard in the pre-Rawls, pre-Nozick era, a rather dry time when Oxford-style analytic philosophy was in saddle and meta-analysis dominated moral philosophy. Normative ethics had hardly any place at all, and ethics was generally treated as a kind of interesting, highly specialized area for exploring concepts and some foundational issues.

What it did not deal with was what, in my naivete, I thought moral philosophy was all about: how I should morally live my own life and my life with others. I once asked at a conference how my professor of moral philosophy, who taught us all about deontology and utilitarianism, related the pacifism of his personal Quakerism to the moral theories he taught us. He reacted in a cool and dismissive way: "I think that question is out of order," he responded, and turned to the next questioner. Another professor told me that, in effect, my kind of questions were old-fashioned. Philosophy, he said, "is a kind of game" that, like chess, "some of us like to play."

I decided that I did not want to spend my philosophical life playing that game and was soon drawn to medical problems as a fine area to make use of my philosophical skills. But I could not do so in any clean philosophical way. As an undergraduate I had a great deal of history, cultural analysis, social science and literature, and a number of friends at Harvard in other fields. I found it impossible to think that, as many philosophers seemed to believe, that moral philosophy comes down to nothing more than good arguments, pure rationality in action. The world, I believe, is more complicated than that. My emerging interest in medicine made it clear to me that the moral problems of that field required that one take responsibility for one's ideas and writings, that ideas do indeed have consequences. It is not a game.

Philosophy has its own biases, is influenced by the culture and regnant ideologies, and is as open to unreason and irrationality as most other forms of intellectual life; it just disguises them better. But philosophers are not trained to notice that. Philosophy can be helpful as part of a cluster of disciplines pertinent to medicine and health care. But moral philosophy as practiced in the Anglo-American mode should never be given a primacy of place. It too often lacks moral sensitivity and self-reflection. It is, so to speak, better able to master musical theory than learn how to carry a tune.

4. What do you see as the most interesting criticism against your own position in philosophy of medicine?

Since those writings of mine that have tried to work philosophy of medicine into my work on health policy have mostly been ignored, or at least put to one side–as if out of the mainstream–I can't recall any "interesting" criticism. What little criticism there has been has been has more or less defended the idea of progress, and many worries have been expressed that my skepticism about the [absolute] value of technological innovation could itself stifle innovation—and one would never want to do that! Yes, that might happen but–given the tight correlation of that innovation and rising costs–it has to be stifled. If those criticisms were not particularly "interesting," they do continually force me to face up to the fact that even a slowing down of innovation would mean the loss of some lives and some improved health. I have found no happy way out of that dilemma.

Rising costs threaten health care systems as a whole. Cutting or downgrading innovation would help systems but at the expense of individuals, an old and familiar policy problem. Actually, come to think of it, the most interesting exchanges I have had are with those who, wanting to avoid pain, invent fanciful scenarios that will take care of the cost problem: greater health care efficiency, medical breakthroughs that will rid us of annoying diseases, and (most delightfully) those who argue that it does matter how much we spend on health care (what could be a better investment?). Just as ancient medicine stressed the value of instilling hope in the ill, even if dying, the modern version is to instill hope in painless cures for cost escalation.

5. How can the most important problems concerning Philosophy of Medicine be identified and explored?

I can envision no formulaic answer to this question. Identifying problems in any old and well-established field requires breadth of vision and an active imagination to turn up something new. But many of the possibilities are probably right there under our noses. A lack of attention to the idea of medical progress came to my attention over 20 years ago, endless cited and praised but not critically examined. It seemed to be universally held that progress is so self-evidently valuable that no examination is necessary (and I discovered that, if its importance is pushed with any vigor, a charge of ludditism could quickly be leveled).

More generally, I have discovered with my work on medicine and the market, and medical research, that there are many problems, usually treated as economics, that could use the attention of those interested in the philosophy of medicine. What, for instance, might best be made of the idea of "medical necessity," often cited in the US as the appropriate standard for providing medical care, but admitting of no consensus at all about its meaning (so too with "reasonable and necessary" as an alternative phrase for medical necessity). Earlier work was done on the concept of "medical futility," which has engendered no consensus either. In both cases it is necessary to ask what the goals of medicine are, which must include the scope and limits of medical aspirations. One reason "medical necessity" has gained no consensus is that its determination has typically been left solely to doctor-patient judgments. The assumption behind putting decisions about necessity in their hands is that every patient is unique and different, not only in the way their disease(s) manifest themselves physically but also in the judgments patients make about what they.

"Marginal benefit" is another much-cited issue in the economics of health care, but again hard to define because of doctor-patient differences about what counts as marginal and about whether, given individual differences in values, any external judgment can be made. Who is to say that a treatment costing $100,000 with only a 1% chance of success is worth it or not? It may not be worth it to a society that has to pay for it but well worth it to a critically ill patient: a 1% chance is better than a 0% chance.

Other examples could be offered, but my point is that a good place to look is in the literature of health-related disciplines–health economics, public health, epidemiology, demography, medical sociology and medical anthropology. They are filled with concepts to beg for some philosophical analysis and which often enough turn out to be fruitful territory for the philosophy of medicine. More generally, modern medicine continues to develop and change, fueled by technological developments, new research, and changing cultural values concerning health and illness. The philosophy of medicine should not be understood as a static field, but one that looks with interest at the constantly shifting medical scene. There is plenty to do.

3
Arthur L. Caplan

Robert and Emanuel Hart Professor of Bioethics, Director Center for Bioethics

University of Pennsylvania

---

1. Why were you initially drawn to Philosophy of Medicine?

My PhD work was in the philosophy of science/philosophy of biology (late 1970s). After doing some work in bioethics as a grad student it was obvious to me that far too little attention had been paid to the philosophy of medicine. There were some writings in the area and especially some provocative European contributions, but philosophers of science had more or less completely ignored medicine—perhaps because it was seen as to practical or applied to qualify as a science.

I was particularly troubled that the dispute about whether or not mental illnesses existed had not received much attention. The whole area of defining diseases, their metaphysical status and how thinking about disease and illness had evolved seem to me to be a fascinating mix of evidence, values and cultural norms. This led me to work with Tris Englehardt and JJ McCartney to put together our reader on health and disease– Concepts of Health and Disease: Interdisciplinary Perspectives Addison-Wesley, 1981. This book did stimulate more attention with philosophy and medicine to the conceptual analysis of health and disease but did not really trigger interest in the broader epistemological and ontological issues that should define the philosophy of medicine.

2. What does your work reveal about Philosophy of Medicine that other academics, citizens, or economists typically fail to appreciate?

This is a big issue but one way to summarize my view is that there is much emphasis these days on evidence-based medicine and outcomes research. However, few within medicine and even fewer medical students and others working in health care have an appreciation for what makes something 'evidence' in medicine!

Similarly while there is much discussion of the ethical challenge of conflicts of interest, there is not much attention being paid to the adequacy of methods and modes of analysis in medicine to control for bias and distortion in findings. Philosophy of medicine could and should speak to this matter but has not had much to say outside of a few recent articles by me, Psasty at Washington University and a few others

3. What, if any, practical and/or social-political obligations follow from studying medicine from a philosophical point of view?

None. One can study medicine and its logics of diagnosis, discovery, test along with nosology, theory change and all the rest and incur no obligation or duty to do anything!!

4. What do you see as the most interesting criticism against your own position in philosophy of medicine?

I am a bit of a positivist and a pragmatist when it comes to medical evidence. But I do think medicine is far more easy to demonstrate as a science then many other 'sciences' since it has the bottom-line of cure or failure to keep it honest. I think many would say this misses much about the social construction power and politics of medicine but I still think there is a core of medicine that rests about theories and claims that are refuted when the patient either dies or gets well.

5. How can the most important problems concerning Philosophy of Medicine be identified and explored?

We need more of a field of philosophy of medicine—more encouragement of philosophers, doctors, nurses, historians to pay attention to philosophical aspects of medicine. I think perhaps the right general reader or textbook might trigger that interest and help create a more vibrant field that would fulfil the function of keeping an eye out for issues and problems.

4

Ruth Chadwick

Director, Cesagen (ESRC Centre for Economic and Social Aspects of Genomics)

Cardiff University

---

1. Why were you initially drawn to Philosophy of Medicine?

I became interested in Philosophy of Medicine quite by chance. I am frequently struck by the ways in which our life course can be affected by a conversation or by a book we happen to pick up. In my case, as a graduate student in the 1970s I came across a collection of essays which included one on the difference between negative and positive eugenics (Roberts, 1972), about which I knew little at the time, and it sparked my interest. Since those days the study of issues in genetics and genomics has exploded, but at that time it was still relatively rare for philosophers to focus on that field.

The next chance event was that I happened to be sitting, at a lunch at Oxford University, next to Professor Sir Hans Krebs, who won a Nobel prize for his discovery of the Krebs cycle. He was kind enough to ask me about my work and it ended up with him arranging for me to be supervised by a geneticist in addition to my Philosophy supervisor (Jonathan Glover) for my doctoral work. I attended the undergraduate courses in Genetics, went into the lab, and had tutorials with David Roberts at the University. This was life-changing, as it not only gave me a quite different understanding of the field, but also raised questions such as 'What is a gene?' which have been and continue to be at the heart of current debates, as we struggle with misleading presentations in the media, for example, about discoveries of a 'gene for' condition x or y. Although my main interest in the field came from a starting point of Ethics, it has always been clear to me that there is a great deal of conceptual groundwork to be done, including the negative-positive distinction mentioned above.

The distinction between positive and negative is one that arises in a number of areas – the difference between negative and positive duties, and negative and positive utilitarianism, for example, but the question of positive versus negative eugenics has very far reaching implications, including not only questions of social policy and ethics, but also issues concerning the very nature of human beings. In order to discuss post- or trans- humanism, it is necessary to have some idea of what counts as human. The more recent discussions on human enhancement have continued to require considerations of these philosophical questions.

2. What does your work reveal about Philosophy of Medicine that other academics, citizens typically fail to appreciate?

I came into Philosophy via the Classics, having originally gone to University to study Latin and Greek, with a view to being an archaeologist. (Although Archaeology has remained a leisure time interest, I turned my professional focus from the past to the future.) So Plato, Aristotle and the Presocratics were where I started out, including work relevant to health issues: I wrote an early paper on Plato and eugenics (Chadwick, 1990). This critically examined what were at the time being claimed as Plato's feminist credentials because of his position on women as potential Guardians. I argued that his eugenic arguments, when set alongside his remarks about women's capabilities, were not promising in this regard.

I think that coming to Philosophy, including Philosophy of Medicine, from this direction, gives you a training which, with its different emphasis from a background in later European Philosophy, has a lasting influence on your outlook. I came to Kant and Mill, for example, quite a bit later, specialising in Kant's philosophy when I took the B.Phil, my first postgraduate degree, at Oxford. I would not say that there is anything particular in my work that reveals something that others fail to appreciate, but I have been particularly interested in the 'internal good' of medicine (and sub-areas of practice such as genetic counselling), that can set logical limits to how we ought to think and act in relation to it. For example, in thinking about what counts as success in genetic counselling, it cannot simply be provision of choice: there must be some justification for the range of choices on offer which is logically connected to the raison d'être of the field itself. (Chadwick, 1993) This line of thought has influenced my work in professional ethics as well as in Bioethics.

A key example of this has been in examining the professional ethics of scientists (Chadwick, 1995). The practice of science has come under increasing scrutiny by those who work in Science and Technology Studies, for example, in relation to debates about both the construction of knowledge and the values inherent in the scientific process, despite claims of neutrality. This is important for Philosophy of Medicine too because many contemporary debates, especially about the future of medicine, involve issues about scientific research in biomedicine. I am thinking in particular about the possibilities of personalised medicine, which crucially depend on large scale population wide genomic research. More controversially, debates about reproductive cloning and regenerative medicine, together with the increasing emphasis on public engagement, raise the questions about how the research agenda is and should be set. I have argued that attention to the 'internal good' of science could be helpful in addressing these issues.

3. What, if any, practical and/or social-political obligations follow from studying medicine from a philosophical point of view?

Bernard Williams was of course interested in the line of argument I have mentioned, and I have found it particularly important. (Williams, 1973) The point about considering what are the goods internally/logically associated with medicine, as opposed to those externally or contingently associated with it, is that it makes us see certain political rhetoric, in particular, in a different light. Medicine, as a whole, and not only genetic counselling in particular, is not logically associated with 'choice', however much that term might carry political advantage in debate. It is about delivering health care to those who need it.

The rhetoric of choice seems to have pervaded every sphere of life but we can see from the debates about health care reform in the United States how an extreme position on this can regard an attempt to extend healthcare coverage as a attack on freedom. It has been prominent not only in debates about reproductive technologies and the right to a child (Chadwick, 1987) but has more recently been central to debates about food and eating habits, as societies increasingly confront the challenge of obesity (See for example Department of Health Choosing Health: Making Healthy Choices Easier Department of Health, London, 2004).

4. What do you see as the most interesting criticism against your own position in philosophy of medicine?

Criticisms might come not only from within Philosophy but also from other disciplines, such as Social Science. The position about goods being logically associated with a particular activity might be attacked from a point of view arguing that it actually depends on context. This would in effect be an argument that the goods identified are **not** logically connected to the activity in question. For example, it might be suggested that the practice of medicine is subject to change. Indeed that is argued today – that there is a shift from the goal of healing to that of enhancement (Gordijn and Chadwick, 2008). Medicine can have a part to play, on this view, in human perfectibility, in overcoming what used to be seen as the natural decline of old age, for example.

This is interesting because it not only forces us to re-examine precisely what is and what is not a given, but also reveals the importance of other philosophical questions to be addressed, such as what counts as an 'enhancement': can we make any sense of the distinction between therapy and enhancement, or the relation between the concepts of enhancement and improvement? (Chadwick, 2008)

Indeed, it does seem to be true that our very concepts are challenged by developments in science and medicine. We only have to think, for example, of debates over the concept of 'embryo' and how advances in research on the embryo and fetus have made some ethical positions problematic. As Mary Warnock famously pointed out (Warnock, 1985), we cannot discover the truth about the moral status of the embryo, however long we might examine it: a decision has to be taken, which requires justification by argument.

Similar considerations apply to the concept of 'death': developments in science make possible new criteria of death, but the requirement for philosophical argument about its definition continues. It might be argued that, while it is true that the discussion about these concepts must continue, philosophers do not have a monopoly on providing the answer. That potential criticism does not seem to me problematic, precisely because I have argued that science, to give one example, and philosophy interact with each other to make certain positions un*think*able: there are emerging trends not only in Philosophy of Medicine but also in Ethics, too, which have formed the focus of some of my recent work (Knoppers and Chadwick, 2005). Bioethics, in particular, has developed relatively recently as a field of study – its development is normally traced to the 1960s. In the first decades it was, for reasons that have been documented, focused to a large extent on the individual patient, although issues in public health have recently attracted greater attention. The extent to which doctrines such as that of informed consent are applicable in relation to population research has come under scrutiny, and it is a matter of debate as to whether new principles are needed, or an emphasis on different principles, or again reinterpretation of existing principles. So context *is* important, but not to the extent that a full-blown relativism is indicated. Even though there might be changes in scope, I would suggest that there is a central core that must be logically associated with the practice of medicine, but there is much room for debate over where the relevant lines are to be drawn.

5. With respect to present and future inquiry, how can the most important problems concerning Philosophy of Medicine be identified and explored?

The identification of problems is best done through dialogue, it seems to me, and dialogue of a multidisciplinary kind. I have already pointed to the ways in which Philosophy has to have regard to developments in science. Social science, however, is also important. It has become obvious in Medical Ethics, for example, that one of the most important tasks of Ethics is to draw attention to the moral dimensions of a situation that might simply be overlooked from specific points of view. Work on public engagement can highlight moral concerns that people have which might be overlooked by a professional philosopher addressing the issues from a specific viewpoint. It is therefore essential to take into account work that elucidates these concerns – and also those of professionals in the relevant fields. This does not mean that social science simply provides data for philosophers to think about. There is also room here for theoretical exchange. We always need to pay attention to the ways in which questions are framed, which might be different within and outside a discipline, which is why working with people from other disciplines can be so fruitful, if challenging. (Chadwick and Levitt, 1997)

References

Chadwick, R. (1992) Ethics, Reproduction and Genetic Control 2nd edn., Routledge, London

Chadwick, R. (1990) 'Feminism and eugenics: the politics of reproduction in Plato's Republic', in A.Loizou and H.Lesser (eds) Polis and Politics: Essays in Greek Moral and Political Philosophy. Avebury, Aldershot

Chadwick, R. (1993) 'What counts as success in genetic counselling?' Journal of Medical Ethics 19: 43-6

Chadwick, R. (2005) 'Professional ethics and the 'good' of science' Interdisciplinary Science Reviews 30: 247-56

Chadwick, R. (2008) 'Therapy, enhancement and improvement' in B.Gordijn and R.Chadwick (eds) Medical Enhancement and Posthumanity Springer

Chadwick, R. and Levitt, M. (1997) 'Coplementarity: multidisciplinary research in bioethics', in S.Gindro, R.Bracalenti and E.Mordini (eds) Bioethics Research: Policy, Methods and Strategies European Commission, Brussels (Report EUR17465EN)

Department of Health (2004) Choosing Health: Making Healthy Choices Easier Department of Health, London

Gordijn, B. and Chadwick, R. (eds) (2008) Medical Enhancement and Posthumanity Springer

Knoppers, B.M. and Chadwick, R. (2005) 'Human genetic research: emerging trends in ethics' Nature Reviews Genetics 6: 75-9

Roberts, C., 'Positive eugenics' in James Rachels and Frank A Tillman (eds) Philosophical Issues: A Contemporary Introduction Harper & Row, New York (1972)

Warnock, M. (1985) A Question of Life Blackwell, Oxford

Williams, B. (1973) 'The idea of equality' in B.Williams, Problems of the Self Cambridge University Press

5

Bill (KVM) Fulford

Fellow of St Cross College, Member of the Faculty of Philosophy and Honorary Consultant Psychiatrist

University of Oxford; and Emeritus Professor of Philosophy and Mental Health, University of Warwick.

---

1. Why were you initially drawn to philosophy of medicine?

I have always had hybrid interests in the sciences and the arts and my route into combining these interests in the philosophy of medicine, and particularly in the philosophy of psychiatry, was through a series of inspiring teachers.

At school, although I was in the science stream, the only prizes I achieved were in English – I regularly won both the school poetry prize and the house play competition (as producer). Later, as a medical student at Cambridge University, my interest specifically in philosophy was sparked through an introduction to Donald MacKinnon, at the time Professor of Moral Philosophy, who took me on as an informal pupil. MacKinnon was a wonderful tutor, a Highland Scot who many have likened to a latter-day Dr Johnson. I remember with affection his booming voice as he spoke of Jeremy Bentham, John Stuart Mill and others in our tutorials (that were often held in the Blue Boar pub!)

MacKinnon subsequently introduced me to Mary (now Baroness) Warnock with whom I was later to do a DPhil. This became possible when, after a period in laboratory research and then qualifying as a psychiatrist through the Institute of Psychiatry in London, I was appointed to a Clinical Lecturer post in Oxford. Again, I was remarkably fortunate in being supported in what was at the time an unusual choice of research interests for a psychiatrist, first by Professor Sir Dennis Hill in London, and then by my Head of Department in Oxford, Professor Michael Gelder. Both pointed out to me the hazards for a young doctor in moving away from laboratory research (I had previously worked in Immunology). But both supported me strongly when I came up with a credible proposal for taking my interest in philosophy forward.

I had similar good fortune on the philosophy side. Mary Warnock arranged for me to have extended periods of supervision with both my philosophical heroes, first with RM Hare and then with Geoffrey Warnock. Having no first degree in philosophy (entry to the Oxford DPhil could be on the basis of written work), this gave me a unique opportunity to develop my knowledge and skills in the particular area of philosophy that I had by then become particularly interested in, a branch of linguistic analytic philosophy called philosophical value theory. As I describe further below, Hare and Warnock (Geoffrey) represented opposite schools of thought in this area (they were also very different personalities) and I built directly on the debate between them in writing my first book, applying philosophical value theory to concepts of disorder, Moral Theory and Medical Practice, which was published in 1989.

Since then there has been a remarkable explosion of cross-disciplinary work between philosophy and psychiatry. Suddenly the new field is out of the closet! It is impossible here to do justice to the tremendously wide variety of colleagues, from both the practice side (including patients and carers as well as professionals) and the philosophy side, who have made vital contributions to developing the new field. But I would like to give a brief overview of developments in this area by way of background to the points that I want to make later.

In less than two decades, then, the philosophy of psychiatry has produced over 40 new scholarly and practice-based groups around the world, coordinated through the International Network for Philosophy and Psychiatry (INPP www.inpponline.org) that was launched from Cape Town in 2002, and including special sections in the two major international psychiatric organisations, the World Psychiatric Association and the European Psychiatric Association. There have been new journals, including Philosophy, Psychiatry, & Psychology which is now in its seventeenth year, and, an important new arrival, the Journal for Philosophy and Psychiatry as an online journal featuring articles in languages other then English. Alongside academic journals, a number of book series have been launched, including International Perspectives in Philosophy and Psychiatry from Oxford University Press. Several new professorial 'chairs' have been established, with associated teaching and research programmes, mostly in Europe though with promising developments in Africa and Asia. The discipline has also attracted considerable research funding; notable grants include £1m for a programme at the University of Central Lancashire, and a substantial DPhil scholarship in Oxford. Last, though certainly not least, the philosophy of psychiatry has already started to have an important impact on policy and practice in mental healthcare. It is to these philosophy-into-practice initiatives, and to their significance for the philosophy of psychiatry as a part of the philosophy of medicine, that I turn in the next section.

2. What does your work reveal about Philosophy of medicine that other academics, citizens, or economists typically fail to appreciate?

Perhaps the main thing that is shown by the remarkable explosion of work in the philosophy of psychiatry over the last two decades, is the importance, both academic and practical, of mental health for the rest of healthcare. Since my own work in the philosophy of psychiatry over this period has been so intimately linked with that of a wide variety of colleagues, I would like to illustrate this point by reference to developments, first academic and then practical, across the field as a whole.

First, then, the importance of philosophy of psychiatry as an academic discipline. This is all about stigma. Sadly, through much of the twentieth century, those working in psychiatry, no less than those actually suffering from mental distress and disorder, were widely stigmatised both within medicine and by the wider public. Philosophy in recent years has suffered something of the same fate, at least in the UK – I was once introduced as a speaker at a medical conference as being a 'philosopher and psychiatrist...' who '...has thus successfully combined two degenerate subjects in one!' The comment was meant (and was taken) as a joke. But there was a serious point behind it for an audience of doctors and medical scientists among whom psychiatry and philosophy were perceived, equally, as being at best lacking a rigorous foundation, at worst failing to produce substantive outputs. That was many years ago. But the biggest challenge that everyone concerned with mental health still faces, whether as a patient or practitioner or policy maker, is the continuing stigmatisation of the discipline as an 'also ran' to other apparently more scientific areas of healthcare such as cardiology and surgery.

Well, the new philosophy of psychiatry gives the lie to all that. In the first place, far from lacking a rigorous foundation, the discipline is firmly rooted in the most rigorous of a whole range of both conceptual and empirical methods. My own background is in analytic philosophy, nothing if not a rigorous discipline, and a discipline that has made a series of important contributions to understanding the conceptual problems underpinning both research and practice in mental health. No less important, however, have been the many contributions from the equally rigorous if more text-based disciplines of Continental philosophy: this includes phenomenology, of course, a discipline that was important at the birth of modern scientific psychiatry at the beginning of the twentieth century, particularly in the work of Karl Jaspers; but it also includes such related disciplines as hermeneutics, existentialism and, more recently, discursive philosophy. The new philosophy of psychiatry has also encompassed innovative combined-methods studies, that is studies combining conceptual with a range of empirical methods. Then again, as to substantive outputs, the second limb of the 'also ran' stigmatisation, these disciplines, philosophical and empirical, have between them produced a whole series of new insights across five principle areas, conceptual analysis, the history of ideas, the philosophy of science, philosophical value theory and ethics, and the philosophy of mind. (See bibliography.)

A first anti-stigma message, then, from the new philosophy of psychiatry, is that mental health, although indeed a particularly difficult area of healthcare, has a robust research infrastructure based on rigorous, and at the same time highly productive, conceptual as well as empirical methods. In this respect, the philosophy of psychiatry thus stands academically, no less than equal with other areas of the philosophy of medicine. There is, however, a second anti-stigma message from the new philosophy of psychiatry, that in at least some areas of conceptual research, the philosophy of psychiatry is actually leading the field.

This is no doubt a more contentious anti-stigma message. 'Psychiatry first, indeed!' someone may be saying. But 'psychiatry first', in some areas at least, is a crucial message to take from philosophy if we are to counter twentieth century stereotypes. The argument runs thus. Until recently, and perhaps still among many of those working in the philosophy of medicine, it has been assumed that the 'big questions' were all to be tackled within so-called mainstream medicine. It was recognised that there were important philosophical questions in psychiatry but it was assumed (usually tacitly) that these questions could be answered by extension from the corresponding answers derived in mainstream medicine. This 'psychiatry second' approach in the philosophy of medicine thus followed closely the stigmatisation of mental health as a second class citizen generally within medicine. Yet the reality has been the reverse of this. Far from the philosophy of psychiatry trading on philosophical work in other areas of medicine, the philosophy of psychiatry has produced a number of insights that could be crucially important for the rest of medicine.

The 'psychiatry first' rather than 'psychiatry second' anti-stigma message is particularly clearly seen when we move from the academic research base of the new philosophy of psychiatry to its growing impact on day-to-day practice. Ethics is a case in point. Bioethics, as it has come to be called, continues to focus its efforts, in relation to core topics such as autonomy of patient choice, in scientifically high-profile areas of medicine such as assisted fertility, organ transplantation and 'end of life' issues. In much of this literature, the issues are discussed against a background of assumed rationality. In psychiatry, by contrast, it is the very concept of rationality that lies behind such highly contested areas as involuntary psychiatric treatment. And work in the philosophy of psychiatry in this area, developing what has become known as values-based practice, has key implications for policy and practice around issues of autonomy across medicine as a whole.

Without going into the details, the essence of values-based practice (vbp) is that it offers a skills-based process for working with complex and conflicting values in medical decision-making that is a partner to the processes offered by evidence-based practice (ebp) for working with complex and conflicting evidence. The theory behind values-based practice was set out in my Moral Theory and Medical Practice, building (as noted earlier) on philosophical value theory; and there have been key contributions also from others working in analytic philosophy, such as John Sadler in the States, in phenomenology, notably the Italian psychiatrist, Giovanni Stanghellini, and using combined philosophical-empirical methods, including Anthony Colombo in the UK, Werdie van Staden in South Africa, and others. But the real excitement of values-based practice, as I describe further in the next section, has been the extraordinarily rapid way in which this philosophically-derived approach has been taken up in policy, training and service developments in mental health. This work, like the academic work on which it builds, has been supported by many colleagues, including clinicians, but also patients and carers and policy makers, in the UK and internationally. And what is happening now, to come to the 'psychiatry first' crux, is that, although developed initially in mental health, the importance of 'vbp' alongside 'ebp' is already becoming increasingly widely recognised across medicine generally. The UK's Royal College of General Practitioners (i.e. family doctors), for example, now has a major curriculum statement on values-based practice and clinical ethics on its website; and Warwick Medical School, a new medical school in the UK, has established a series of initiatives in this field with the aim, ultimately, of creating the first fully values-based as well as evidence-based undergraduate and postgraduate medical curriculum.

Work in the new philosophy of psychiatry, then, both academic and practical, is set to reverse twentieth century stereotypes: mental health, as a medical discipline, has a rigorous research base, conceptual as well as empirical; in its conceptual research it is leading the field at least in some areas; and building on this research, it is leading the field, too, with policy-into-practice initiatives in values-based practice.

It is worth adding that these 'psychiatry first' messages from the new philosophy of psychiatry were anticipated, in broad outline, by one of the key twentieth-century exemplars of linguistic analytic or 'ordinary language' philosophy in the Oxford tradition, JL Austin. Austin, as White's Professor of Moral Philosophy in Oxford in the period immediately following the Second World War, was perhaps the most explicit advocate of philosophers focusing on ordinary language as a resource for their work. Of course Austin never suggested (as he has often been accused of suggesting) that ordinary language is the only resource needed by philosophy. But he did argue that careful attention to ordinary language, supplemented where appropriate by relevant empirical methods, was, as he famously put it 'one way of 'getting started' with philosophical problems'; and at the end of his most explicitly methodological paper on this topic ('A Plea for Excuses'), he pointed to what we would now call psychopathology as a rich resource of ordinary language for philosophers to study. The master of epigrams, he concluded 'there is gold in them thar hills', thus directly anticipating the subsequent explosion of work in the new philosophy of psychiatry.

3. What, if any, practical and/or social-political obligations follow from studying medicine from a philosophical point of view?

Taking this question, again, specifically from the perspective of philosophy of psychiatry, will suggest implications for the philosophy of medicine as a whole. To anticipate a little, experience in the philosophy of psychiatry shows that bringing philosophy fully into medicine alongside the empirical sciences, is the key to re-establishing medicine as a person-centred as well as research-led discipline. Strictly speaking, this is not an 'obligation' that follows from studying medical philosophy, as the above question puts it. It is, rather, an enriched resource for fulfilling the obligations of medicine as a humanitarian discipline.

Values-based practice provides a case in point of this enriched resource. Bioethics, as it originally arose in the second half of the twentieth-century, was a response to the increasingly difficult ethical issues raised by technological and scientific advances in medicine. There had of course always been 'medical ethics' (the Hippocratic 'Oath' goes back to classical times) but it was the scientific progress of medicine, and the increasingly urgent ethical and other values-related issues that this raised, that drove the need for a more sophisticated ethical infrastructure. But in recent years, bioethics has become increasingly impoverished as a practical discipline as it has shifted progressively away from its roots in philosophical research and towards a more legalistic paradigm. As an academic discipline, of course, bioethics remains richly connected to philosophy. But as a practical discipline, i.e. in the applications of bioethics in practice, it has become increasingly an extension of medical law.

This is understandable up to a point. Policy makers and practitioners alike, faced with the conflicting values that underpin so many medical ethical issues, look for guidance on what they should actually do rather than for nuanced philosophical discussion. Such guidance, in the form of quasi-legal codes of practice backed by medical law, does indeed have an important place in framing day-to-day decision-making. But no set of legal and/or ethical rules can ever be sufficiently complete to replace the judgements that are required in the full complexity of individual clinical cases. Worse than that, the attempt to rely merely on a set of rules ends up being highly counter-productive ethically. As the British social scientist, Priscilla Alderson, pointed out many years ago, there is a paradox here: bioethics emerged to strengthen the autonomy of choice of patients in a healthcare context in which they felt disenfranchised by the expert scientific knowledge of professionals; and now the emergence of a new 'professionalised ethics' had left them doubly disenfranchised.

Values-based practice, then, is (in part) a response to the increasing legalisation of ethics. Codes and ethical regulation, it is important to emphasise, provide a valuable framework for decision-making. But work in philosophical value theory suggests that this framework is best understood as a framework of values within which balanced judgements have to be made by individual patients and individual practitioners working together in partnership in the particular circumstances of each individual situation. This model was originally suggested by HLA Hart, at the time Professor of Jurisprudence in Oxford and also a practicing barrister; and it has more recently been applied to the Human Rights Act by one of the UK's most senior lawyers, Lord Woolf. Interestingly, it is also implicit in one of the foundational (and much abused) approaches to medical ethics, Tom Beauchamp and James Childress' Principles of Biomedical Ethics. As prima facie principles, Beauchamp and Childress' famous four principles provide a framework (no more) for medical decision-making. Beauchamp and Childress make this crystal clear in their book. But it has taken a whole programme of further philosophical research, dipping back into philosophy (into philosophical value theory, phenomenology, etc, as above) to develop the skills-base and other processes of values-based practice required for balanced decision-making in individual cases.

So, exactly what are the practical developments in values-based practice? Well, first, we now have a whole raft of training materials for values-based practice. The first training manual, 'Whose Values?', was developed and piloted with frontline mental health staff and patient organisations, in a partnership between the Sainsbury Centre for Mental Health and the Philosophy and Ethics of Mental Health programme at the Warwick University Medical School. The Sainsbury Centre is one of the UK's largest mental health NGOs and the successful development of the manual depended critically on the input of a number of very experienced trainers, including Kim Woodbridge (first author of the manual), Malcolm King and Toby Williamson. Whose Values? was launched by the then Minister of State in the UK's Department of Health with responsibility for mental health, Rosie Winterton, and materials from the manual have subsequently been included in a variety of training materials covering more specific areas of policy and practice: these include web-based materials for e-learning to support a skills training programme for multidisciplinary teams; and a series of workbooks to support implementation of the UK's new Mental Health Act (the legislative framework for involuntary psychiatric treatment). The latter training materials, for the new Mental Health Act, are distinctive in being not only values-based but also based on evidence derived particularly from patient-led research (particularly by Sarah Dewey). There are also training developments in a number of other European countries and at the University of Pretoria in South Africa.

Second, and no less important to the infrastructure for values-based practice, has been a series of policy initiatives. Thus, values-based practice was originally introduced into the work of the UK's Department of Health through a joint programme between patients and professionals that led to the adoption of a National Framework of Values for the UK's National Institute for Mental Health in England (NIMHE). NIMHE wat at the time the section of the UK's Department of Health that had responsibility for delivering on government targets for mental health as defined by a key policy document, the National Service Framework for Mental Health. The NIMHE Framework of Values thus provided a strong policy platform for ensuring that values-based as well as evidence-based approaches underpin service development in all areas of mental health and social care. Correspondingly, values-based practice has now been adopted as one of the two key themes (the other being evidence-based practice) underpinning a national initiative in generic skills training, the Ten Essential Shared Capabilities (the '10 ESCs'). The '10 ESCs' in turn underpins new ways of working for psychiatrists and others that are more patient-centred and multidisciplinary in approach; and these new ways of working in their turn underpin a variety of more specific policies concerned, for example, with such areas as recovery practice, delivering race equality, and the role of patients and informal carers as 'experts by experience'. Values-based practice has also been incorporated into wider service commissioning and service audit, for example in the Health Standards for Wales.

Consistently with the focus on individual values and shared decision-making in values-based practice, an important feature of the above training and policy initiatives has been strong partnerships, between patients, professionals and policy makers working together on an equal basis. Partnership has been particularly important in a third area of expanding resources for values-based practice, viz. diagnostic assessment. The work on diagnosis follows a series of international research seminars, initiated by John Sadler at UT Southwestern Medical Center in Dallas, and then continued in London supported by the Department of Health and the World Psychiatric Association, exploring the role of values in psychiatric diagnosis. Consistently with the partnership model, these seminars brought together patients and carers with clinicians, researchers and policy makers. At the last of the seminars, in 2006, which was hosted jointly with the Mental Health and Substance Abuse Section of the World Health Organization, we moved from theory to practice with the launch, again by the Minister, Rosie Winterton, of an extensive programme of consultation and service development in diagnostic assessment. The results of the programme, which was co-led by Laurie Bryant, Lu Duhig and myself (respectively as Service User, Carer and Values-based practice leads for NIMHE), included a wonderful range of examples of innovative practice from the field. These were published recently as 3 Keys to a Shared Approach in Mental Health Assessment (the '3 Keys') and we are now in the early stages of putting together an implementation programme to support wider dissemination of best practice in assessment.

The '3 Keys' programme on assessment has been important in bringing together at the sharp focus of diagnosis, the need noted above to combine the resources of generalised evidence-based science with an approach to individual decision-making that is also fully values-based, and hence responsive to the needs, wishes, strengths and other values of the individual concerned. Diagnosis has been to some extent neglected by bioethics. Ethical issues have been explored in relation to the proper and professional conduct of diagnosis in the bioethical literature. But the actual diagnosis itself, i.e. how a presenting problem is understood, as distinct from what we do about it, has been assumed to be somehow off-limits, in effect reserved to the scientific expertise of the medical professional. The theory of values-based practice suggests, to the contrary, that diagnosis (how a problem is understood) no less than treatment (what we do about it) is a deeply value-laden matter, and hence an area of medical decision-making to which the resources of values-based practice alongside those of evidence-based practice are equally relevant.

The programme on diagnostic assessment brings us back to the issue of stigmatisation. It might be thought, consistently with the stigmatisation of psychiatry as a scientific also-ran, that while it might be true that psychiatric diagnosis is value-laden, surely the diagnoses made in more high-tech areas of medicine are 'purely' scientific (being based on x-rays, blood tests, etc). This thought would be quite wrong, however. Without, again, going into the details, work in philosophical value theory suggests, 1) that the relatively explicit value-ladenness of diagnostic assessment in psychiatry is a direct consequence of the fact that in psychiatry (as against bodily medicine) we are concerned with areas of human experience and behaviour (emotion, desire, belief, motivation, sexuality, etc) in which people's individual values are highly diverse and hence may come into conflict (and hence become noticed) in practice, and 2) that far from reflecting a (supposed) lack of scientific development of psychiatry, the value-ladenness of diagnosis, and with it the need for values-based alongside evidence-based approaches, will actually increase (rather than be reduced) with every advance of science and technology in all areas of medicine. In other words, with scientific and technical advances, the processes of diagnosis in bodily medicine which currently appear value-free, will come increasingly to be, like the processes of diagnosis in psychiatry, overtly value-laden. Why? Well, essentially because scientific advances open up new choices in medicine and with new choices go more diverse individual (and hence potentially conflicting) values – think for example of the diverse individual (and often conflicting) values opened up in reproductive medicine by advances in assisted fertility.

Once again, therefore, the philosophy of psychiatry is leading the field, first (at the level of theory) in showing clearly that diagnosis is very far from being 'off limits' to values in medicine; and second (at the level of practice) in developing the decision-support tools (the tools of values-based practice) that, although currently being applied to diagnostic assessment mainly within mental health, will increasingly be needed across medicine as a whole.

4. What do you see as the most interesting criticism against your own position in philosophy of medicine?

Certainly the most interesting criticism of my position (interesting to me, at any rate) is that it is philosophically old-fashioned. When Moral Theory and Medical Practice came out it was described by a colleague (a philosopher) as a 'philosophical coelacanth'! What this colleague had in mind was that the linguistic analytic school of ordinary language philosophy, on which as I said earlier, Moral

Theory and Medical Practice, and hence the subsequent development of values-based practice, has been based, was at its height over fifty years ago, in Oxford, i.e., in the middle decades of the last century.

Well, I am 'guilty as charged'. Or at any rate, I accept the charge of being old-fashioned, but as a compliment not as a criticism. In the first place, my early exposure to philosophy – through the Warnocks, Hare, MacKinnon and others – made clear to me that it was the approach of the 'Oxford school' that was most directly relevant to the problems of day-to-day practice that I was encountering in my work as a young doctor training in psychiatry. It was, specifically, conceptual problems, problems arising in particular from the concept of mental disorder, and these problems specifically as they raised issues around the relationship between descriptive and evaluative meaning in medicine and psychiatry, that seemed to me to be at the very heart of the challenges, both theoretical and practical, that were presented by psychiatry as a medical discipline. So it was natural that I should be drawn to a philosophical school the focus of which was conceptual analysis, pursued particularly through the careful study of ordinary language (including therefore the language of psychiatry), and with a particular interest in the relationship between descriptive and evaluative meanings.

I mentioned JL Austin earlier as a prime exemplar of the Oxford school. Unfortunately, Austin had died before I got to Oxford but I learnt a great deal about him particularly from Geoffrey Warnock, who had been a pupil of Austin's and was one of his literary executors. Geoffrey Warnock's subsequently published philosophical biography of Austin, called, simply, J L Austin, spells out some of the many ways in which linguistic analysis, and Austin's take on linguistic analysis in particular, has many important lessons for us in developing the philosophy of medicine. One example of these lessons is Austin's emphasis on the importance of what he called philosophical 'field work', i.e. the need for philosophers, working in partnership with empirical scientists and other practitioners, to repeatedly go back to the 'data' of the real world. A second example is his neglected but key idea, that many philosophical problems could fruitfully be pursued, on the model of scientific problems, through a teamwork approach rather than relying solely on the standard philosophical model of the lone researcher. These and other ideas of Austin's have been deeply influential in the development of the new philosophy of psychiatry and, like indeed the coelacanth, they have proved remarkably robust.

That said, there is an important sub-category of the 'coelacanth criticism' that carries, perhaps, rather more weight, namely that I rely too heavily on the 'fact/value' divide. Again, I accept the observation but not as a criticism. True, I have used, perhaps overused, the distinction between descriptive and evaluative meaning in my work; and I have used this distinction, as I noted earlier, particularly as it was explored in a long-running debate between two of my DPhil supervisors, RM Hare and Geoffrey Warnock. The opposed positions of Hare and Warnock on the relationship between descriptive and evaluative meaning produced, as all the great debates in philosophy have produced, a rich resource of insights and ideas for working with descriptive and evaluative meaning. It is this resource that I have applied both in my theoretical work and in developing values-based practice. But exposure to that debate also made clear to me the extent to which, again like all the great debates in philosophy, far from resolving the issues, it deepened them. There is no 'final answer' (as yet or ever likely to be) about the logical relationship, the relationship of meaning, between descriptive and evaluative meanings. But this is not to say that the distinction, and the moves and insights into the distinction developed by philosophers, cannot be put to good use when applied in a practical context to solving real world problems – if we waited for a 'final solution' to the foundations of mathematics, we would never get started!

My own position, for what it's worth, on descriptive and evaluative meaning, is neither with Hare (who believed that they were always logically distinct, a position usually characterised as 'prescriptivism'), nor with Warnock (who, with other 'descriptivists', believed that at least some sets of descriptions entailed evaluative meanings), but closer to that of the American philosopher, Hilary Putnam, in his Collapse of the Fact/Value Dichotomy and Other Essays. What this position amounts to is that there is indeed a useful distinction between descriptive and evaluative meaning but that there is no dichotomy, i.e. that the distinction cannot be drawn in every case or fully, that it does not go 'all the way back'. Interestingly, there is direct (linguistic analytic) evidence in support of this idea from psychopathology, specifically in the forms (descriptive and evaluative) that different delusions may take, as I pointed out in chapter 10 of my Moral Theory and Medical Practice. Again, the 'distinction but not a dichotomy' story is certainly not the last word, theoretically speaking, on the fact/value divide, and the lesson of history is that the debate about descriptive and evaluative meaning will continue. But the successful development of values-based practice is sufficient proof that without coming to any final conclusions, the moves in the debate thus far have considerable practical utility.

5. With respect to present and future inquiry, how can the most important problems concerning Philosophy of Medicine by identified and explored?

Here I can be brief! In a word, I believe the most important problems for the philosophy of medicine can be identified and explored through one or other variant of Austin's philosophical field work. What I mean by this is that the best way for us to identify and explore problems in the philosophy of medicine is by philosophers working closely with practitioners, with patients and carers as well as with clinicians and policy makers, to develop a shared empirical-conceptual agenda. From some purist perspectives, in Continental as well as in analytic philosophy, this might seem like philosophers metaphorically getting their hands dirty with empiricism. But as I noted earlier this was precisely what Austin recommended, including in his notion of philosophical field work the idea of research strategies built on teamwork and employing combined philosophical-empirical methods. I have illustrated the potential of this approach for carrying philosophy into practice with the example of values-based practice. There are many other examples of philosophy-into-practice emerging in the new philosophy of psychiatry: these include work on tacit knowledge and the irreducibility of individual judgement, notably by the Professor of Philosophy and Psychiatry at the University of Central Lancashire, Tim Thornton; new research, both analytic and Continental, at the interface between philosophy and cognitive neuroscience (on delusion, for example, by Martin Davies and others in Oxford, by Louis Sass at Rutgers, and others); the emergence of 'naturalised phenomenology' (for example, in the work of Jean Petitot, Francisco Varela, Shaun Gallagher); and a whole programme of research in political philosophy by Patrick Bracken, Philip Thomas and others, supporting a strong movement for service change, called 'critical psychiatry'.

There are many barriers to successful philosophical field work: research funding, access to major journals, career pathways and so on, are all more challenging for the interdisciplinarian. But perhaps the greatest challenge is of philosophical field work losing its distinctive philosophical-ness, and collapsing to become, merely, field work. The British philosopher, A J Ayer, once remarked that philosophy, unlike the empirical sciences, does not progress by closing down on final conclusion; but neither does it stay still; it moves rather in a spiral, periodically revisiting the deep questions, and, with luck, mining one or more new insights with each iteration. In the philosophy of medicine, we might add that every now and again, and again with luck, one of Ayer's 'new insights', though indeed still partial and incomplete, will be found to have a purchase on some real world problem. This is how it has been with values-based practice. The insights from the Hare/Warnock debate, in particular into the relationship between descriptive and evaluative meanings, together with other philosophical work as outlined above, have supplied the theoretical basis for a move from a centralised (quasi-legal) ethics of healthcare to what amounts to a democracy of values. The danger now, of course, a danger that arises directly from the very success of the new approach, is of ossification around a new orthodoxy. It is this danger that the philosophy of medicine meets head on, not by claiming final conclusions, such would indeed be the route to ossification, but rather by continually spiralling back, in Ayer's image, to mine new insights from the deep questions of general philosophy in an open and dynamic relationship with practice.

References

The Philosophy of Psychiatry

1.1 A brief introduction and overview of the new philosophy of psychiatry, setting it in context with twentieth century developments in its two parent disciplines, is given in Past Improbable, Future Possible: the Renaissance in Philosophy and Psychiatry. Fulford, K. W. M., Morris, K. J., Sadler, J. Z., and Stanghellini, G. (2003) Chapter 1 (pps 1-41) in Fulford, K. W. M., Morris, K. J., Sadler, J. Z., and Stanghellini, G. (eds.) Nature and Narrative: an Introduction to the New Philosophy of Psychiatry. Oxford: Oxford University Press.

1.2 Exemplar publications in the philosophy of psychiatry are to be found in PPP, Philosophy, Psychiatry, & Psychology, (see www.press.jhu.edu/journals/philosophy_psychiatry_and_ psychology/) and other journals (see text); and in various book series, including the Oxford University Press book series, International Perspectives in Philosophy and Psychiatry (IPPP).

http://www.oup.co.uk/searchresults?searchType=UKSimple&
searchStart=1&pathToSearch=&keywords=International+
Perspectives+in+Philosophy+and+Psychiatry&Go.x=20&Go.
y=12.

1.3 Developments in the five main areas of philosophy of psychiatry (conceptual analysis, history of ideas, philosophy of science, ethics and philosophical value theory, and philosophy of mind) are described, with a CD-rom of key readings, in The Oxford Textbook of Philosophy and Psychiatry. Fulford, K.W.M., Thornton, T., and Graham, G. (2006). Oxford: Oxford University Press.

1.4 Details of a Masters-level MA programme, based on the Oxford Textbook of Philosophy and Psychiatry (see above), are given at www.uclan.ac.uk/philosophyandmentalhealth

1.5 Other locally taught and distant learning Masters-level programmes include: MSc in Philosophy of Mental Disorder at King's College, London: http://www.kcl.ac.uk/schools/humanities/depts/philosophy/prospectivegraduate/mscphilmental/

1.6 MA/MSc/Diploma/PGA in Philosophy and Ethics of Mental Health at Warwick University, UK:
http://www2.warwick.ac.uk/fac/med/study/cpd/subject_index/pemh/

1.7 MPhil in Philosophy & Ethics of Mental Health at University of Pretoria, South Africa: www.up.ac.za/pemh

1.8 The DPhil programme in the Faculty of Philosophy, University of Oxford, is advertised at www.philosophy.ox.ac.uk.

1.9 Activities in the philosophy of psychiatry, including international conference announcements, are given at www.inpponline.org.

Theory and empirical base of values-based practice

2.1 The theory of values-based practice is set out in my Moral Theory and Medical Practice (Cambridge: Cambridge University Press, 1989), and a number of subsequent publications (see website below).

2.2 An example of recent important phenomenological work on values is, Stanghellini, G. (2004) Deanimated bodies and disembodied spirits. Essays on the psychopathology of common sense. Oxford: Oxford University Press.

2.3 The linguistic-analytic method is used to explore values in psychiatric diagnostic classification in Sadler, J.Z. (2005) Values and Psychiatric Diagnosis. Oxford: Oxford University Press.

2.4 Examples of combined empirical-analytic research include, 1) 'Evaluating the Influence of Implicit Models of Mental Disorder on Processes of Shared Decision Making within Community-based Multi-disciplinary Teams'. Colombo, A., Bendelow, G., Fulford, K.W.M., and Williams, S. (2003) Social Science & Medicine, 56: 1557-1570; and 2) Van Staden, C.W. and Fulford, K.W.M. (2004) 'Changes in semantic uses of first person pronouns as possible linguistic markers of recovery in psychotherapy'. Australian and New Zealand Journal of Psychiatry, 38:4, 226-232.

2.5 The practical methods used for values-based practice, and for combining values-based with evidence-based approaches, are illustrated with a detailed case history of 'the artist who couldn't see colours', in K.W.M. Fulford (2004) 'Ten Principles of Values-Based Medicine'. Chapter 14 in Radden, J. (ed) The Philosophy of Psychiatry: A Companion. New York: Oxford University Press.

2.6 A new series on values-based practice in medicine is being launched next year by Cambridge University Press. The launch volume is Fulford, KWM., Carroll, H and Peile E, 'Essential Values-Based Practice: linking science with people'.

Practical tools for values-based practice in mental health

"Training"

3.1 'Whose Values?' (Woodbridge, K. and Fulford, K.W.M. (2004) London: The Sainsbury Centre for Mental Health). Available from www.scmh.org.uk

3.2 Care Services Improvement Partnership (CSIP) and the National Institute for Mental Health in England (NIMHE) (2008) Workbook to Support Implementation of the Mental Health Act 1983 as Amended by the Mental Health Act 2007. London: Department of Health.

"Policy"

3.3 NIMHE (2004) The National Framework of Values for Mental Health. Available at www.nimhe.org.uk/ValuesBasedPractise.

Also available in Woodbridge and Fulford (2004) Whose Values?, see below.

"Service Development"

3.4 The Ten Essential Shared Capabilities: A Framework for the Whole of the Mental Health Workforce. Department of Health (2004).

3.5 The National Institute for Mental Health in England (NIMHE) and the Care Services Improvement Partnership (2008) 3 Keys to a Shared Approach in Mental Health Assessment. London: Department of Health.

www.3keys.org.uk/downloads/3keys.pdf

Valuse-based practice in other areas of medicine

4.1 The Royal College of General Practitioners (2005) curriculum statement is: Curriculum Statement: Ethics and Values Based Medicine at
www.rcgp.org.uk/gpcurriculum/pdfs/ethicsAndVBPsfRCGP CouncilDec2005.pdf.

4.2 An important statement of the close links between values and evidence in all areas of medical decision-making, including diagnosis, is given in the introduction to Sackett, D.L., et al.'s (2000) Evidence-Based Medicine: How to Practice and Teach EBM (2nd Edition). Edinburgh and London: Churchill Livingstone.

4.3 A comprehensive annotated bibliography on values-based practice is given in the Warwick University Medical School website http://www2.warwick.ac.uk/fac/med/study/cpd/subject_index/ pemh/vbp_introduction.

6

Henk ten Have

Director and Professor
Center for Healthcare Ethics, Duquesne University, Pittsburgh

---

1. Why philosophy of medicine?

When I started studying medicine in 1969 many controversies existed about university education in general and medical education in particular. Just after the students' movement in France, Dutch students also demanded to be more involved in university policies. They organized protests and manifestations for various reasons, but particularly because the government proposed to increase tuition fees while university education at that time started to become less elitist. In Leiden medical school, the curriculum has just been reformed with less emphasis on the natural sciences. For the students, however, the reform was not sufficient; they wanted more teaching in social and psychological subjects as well as practical exercises earlier in the curriculum to show the relevancy of the courses for later medical practice. In retrospect what we experienced in this first year of medical education was disappointment (we learned later that this has been described by sociologists as the institutional development of cynicism). Most of us had chosen to enter medicine out of idealistic motives; we wanted to care for other human beings although the profession was certainly not the best paid one and the educational career was exceptionally long. The personal idealism of helping people was however confronted with impersonal structures, procedural approaches and psychological detachment. The first year of the curriculum focused primarily on statistics, physics, chemistry, biology. The very intensive program of courses did not provide the impression that medicine has anything to do with human beings in specific conditions and differing social and cultural settings. Activist students protested in order to put pressure on the faculty to revise the curriculum more substantially. But it was also clear to some of them that in fact an overall conception of medicine and its purposes, a more philosophical approach, was missing. Around the same time, the reputation of philosophy professors was spreading among the students. Quite a number of medical students decided just for curiosity to attend the public lectures of two of the most famous professors. In fact they involved themselves in an alternative rather than 'hidden' curriculum. Every friday afternoon in the ancient academy building of Leiden University, Professor Gabriel Nuchelmans, specialist in analytic philosophy, logic and medieval philosophy was teaching two courses on different subjects. His courses were open for anyone, and over the years a large audience has been gathering, not only interested in the subject but also because he was an excellent teacher. The other professor attracting many students was Cees van Peursen, an energetic, stimulating teacher with a very broad range of interests. Epistemology and philosophical anthropology were his main areas, but he also made explorations in metaphysics and philosophy of culture. His public lectures were focused on particular philosophers (such as Levinas or Spinoza) or philosophical schools (such as existentialism); they were connected with intensive reading sessions during which a small group of students was slowly reading a major volume. These exceptional teachers created a real interest in philosophy among their students. At the same time, it was obvious that they did not address fundamental questions regarding medicine and health care, at least not in a direct way. Philosophical reflection, however, is at the end aimed at elucidating basic queries concerning human existence. The philosophical courses demonstrated that science is not a mechanical activity of reproducing knowledge but a critical, analytical and reflective challenge to the intellect. Medical students all over the country decided to combine the study of philosophy with that of medicine. When later, in 1982, the initiative was taken to establish the Dutch Society for Philosophy and Medicine, it turns out that there were approximately one hundred interested persons with a double degree.

Another professor of philosophy who motivated me to go specifically into the direction of philosophy of medicine was Marius Jeuken. Biologist, philosopher and Jesuit, he was teaching philosophy of biology in the Science Faculty. When I had to choose the main subject for my master in philosophy, philosophy of medicine had not yet attracted much attention. The discipline closest to it was philosophy of biology. Jeuken addressed many relevant issues such as body and mind, science and values, nature and life, freedom and determinism. He also directed the Institute of Theoretical Biology with excellent library and study facilities. He encouraged me to address the interrelations between medicine and philosophy. As a result, in one of my first publications, published by the leading Dutch journal of philosophy, I argued that philosophy of medicine should be considered as a legitimate philosophical discipline (ten Have, 1980).

Under Jeuken's guidance I studied what philosophers like Plato, Aristotle and Kant have said about medicine or medical subjects. More or less by accident I discovered that Jeremy Bentham has had a considerable influence on contemporary medicine. He was one of the first to argue that there is a need for activities in public health and for deliberate health policy. He emphasized the importance of prevention and the utility of creating a separate Ministry of Health. As a philosopher, Bentham inspired the emergence of the sanitarian movement with disciples such as Thomas Southwood Smith and Edwin Chadwick. It was a perfect example of how philosophy and medicine have been interacting in the past. For Bentham, medicine was the leading model for reformation and modernization of society. What the physician is doing at an individual level, the lawmakers should do at the level of the 'political body'; they should address the 'political nosology' on the basis of experience, observation and experiment. On the other hand, his focus on social conditions promoted the view that major health problems such as epidemic diseases are not the result of individual behavior or contamination, but rather the sequel of poor and unhygienic living conditions due to the Industrial Revolution with overcrowding, bad air, pollution and insalubrious drinking water. In his view, individual patients should not be the target of medical intervention but the environment and social conditions as main sources of diseases.

My master thesis on the influence of Bentham on contemporary medicine and health care was later elaborated into a PhD dissertation (ten Have, 1983). The historic example however was located in the wider context of interaction between medicine and philosophy. The two disciplines have for most of their history been intimately related. Only recently, with the emergence of medicine as a natural science, they separated. The warning of the Swiss psychiatrist Eugen Bleuler (1921) that medicine and philosophy should be kept apart, otherwise one will end up with a mixture of chocolate and garlic, has been taken seriously. Medicine is natural science; philosophical speculation is not only useless but also dangerous for its development. Philosophy has only produced a graveyard of dead systems of ideas.

2. Re-appreciation of philosophy of medicine?

What started as a personal quest for philosophical reflection on medicine transformed rapidly, at different places, into a systematic and disciplinary exploration of a new field of scientific enquiry: philosophy of medicine. The appearance of specialized journals such as The Journal of Medicine and Philosophy (since 1976) and Metamed (1977; renamed Metamedicine since 1980; Theoretical Medicine since 1983) as well as the book series Philosophy and Medicine (since 1975) showed that in many countries in more or less the same period of time philosophical reflection on medicine has been emerging. But, as in other countries, there have been two streams in the development of medical philosophy: medical ethics practiced by theologians and moral philosophers, and philosophical studies focused on concepts, theories and methodologies of medicine, by physicians. Only in the mid 1980s these streams flowed together with a growing dominance of the ethical one. In my scientific work I have always tried to bring the increasing focus on bioethics back to the initial and fundamental context of philosophy of medicine. In fact, the intellectual movement towards bioethics was facilitated by two reductions. The notion of 'bioethics', introduced in 1970 has been reduced from its initial wide scope proposed by Potter to a more limited one, focused on an enlarged version of medical ethics. Furthermore, the fundamental and critical debate about health care was reduced to a discussion of normative issues, and primarily to the question what individuals ought to decide. Bioethics became 'pars pro toto' for the wider movement of philosophical reflection. These two reductions imply that important concerns and questions have disappeared from the agenda of bioethical debate. In this respect it is important to clarify the context of philosophy of medicine in which bioethics has been born.

A. Context of emergence of philosophy of medicine

Critical reflection was closely linked to the progress of medical science in the 1950s and 1960s. In the Dutch context the evolution of philosophy of medicine was related to the tradition of anthropological medicine and general practice, that both requested a focus on the patient as a whole person and thus a more holistic methodology, as well as the tradition of history of medicine (ten

Have and van der Arend, 1985). The first phase of criticism was characterized by the identification and critique of 'medical power'. Particularly influential was a booklet of Van den Berg (1969) scandalizing the unprecedented power of medicine. Within the medical discourse itself, it was argued that the role and efficacy of medicine is often overestimated (McKeown, 1976).

The second phase focused specifically on the negative impact of medical power. This power, exercised by professionals, is often associated with arrogant and paternalistic behavior as well as a tendency to expansion into other areas of human and social life. The concepts of 'medicalisation' and 'iatrogenesis', introduced by Zola (1973) and Illich (1975), appeared particularly fruitful. Many studies emphasized the interrelation between medicine, society and culture, and they could be easily connected with philosophical criticism from structuralism (Foucault) and critical theory (Adorno, Habermas). Such critique led to the third phase of counterbalancing medical power, either by imposing limits through legislation and emphasis on patient rights, or by articulating individual autonomy and decision-making, or by creating and utilizing alternative systems such as complementary, holistic, and humanistic medicine and self-help (Aghina, 1978, van Dijk, 1978; ten Have, 1980).

The various activities of critically analyzing present-day health care promoted a broader examination of the presuppositions, foundations, methods, concepts and values of modern medicine. The usual approaches and assumptions could no longer be taken for granted. Rethinking the philosophy of medicine was apparently motivated by the need to clarify the image of the human being, not only presupposed in medical activity but also stimulated and reinforced by medical science. As philosophy in action, medicine tries to remake man and reality (Pellegrino, 1976; Engelhardt, 1974). Medicine always proceeds with certain implicit ideas about what human beings are and should be. This motivation is connected with social movements to 'humanize' medicine and to make it more 'holistic'.

Another motivation for philosophical analysis was the need to clarify the scientific character of medicine, reflecting on the methods of clinical judgment and clinical decision-making (Wulff, 1980). Elucidating the so-called medical model was particularly imperative in discussions with protagonists of anti-psychiatry and alternative medicine. Kuhn's paradigm theory of scientific development was applied to medicine (Verbrugh, 1978), to determine the demarcation between science and non-science but also to discover the special nature of general practice and family medicine. In connection to medical methodology, basic concepts such as health, disease, illness, normality, complaint and suffering were intensively discussed (ten Have, 1984). A third motivation to engage in philosophy of medicine was created by the moral problems of medical progress. It is not simply that there are new or different problems but medical ethics itself is in a crisis. A first proposal to redefine medical ethics, based on the concept of human dignity and on a broader image of man as a relational being, was published by Paul Sporken (1969) who was also appointed as the first professor of medical ethics in a medical school (in 1974 in Maastricht). During the 1980s bioethics rapidly institutionalized; all eight medical schools in the country created chairs and specific departments or centers in medical or bioethics with ethics teaching as mandatory component of the curriculum.

B. From internal to external morality

The rapid emergence of bioethics can be explained as the transformation from internal to external morality in the area of health care. Traditionally, medical ethics referred to the deontology of the medical profession. Deontology expressed the internal morality of medicine, i.e. the specific values, norms and rules intrinsic to the actual practice of medical care. They define implicitly good clinical practice and determine what a good professional is. The growing power of postwar medicine discredited this notion of internal morality and replaced it rapidly with values, rules and norms external of medicine. These normative determinants were prevailing in social, cultural and religious traditions that influence the context in which medicine is practiced. There was increasing consensus that these determinants should be more important in the regulation of medical practice than the usual internal ones. The new bioethics emphasized primarily external morality, with moral notions such as 'individual autonomy' and 'social justice' (ten Have, 2001).

This shift from internal to external morality had significant consequences. It has been instrumental in creating a distinct profession of bioethicists, but it also promoted the development of health care legislation (in some countries like France specifically labeled as 'bioethics laws'), separate institutions (bioethics committees) and educational and training programs (masters in bioethics). However, the primary emphasis on external morality also encouraged a particular view of medicine as a neutral transaction, an enterprise, even trade or business, aimed at exchanging technological assistance and expertise with the demands and needs of autonomous persons (clients or consumers). By developing into an autonomous discipline assisting health practices, bioethics has at the same time become a component of the technological order. It has been dominated by an engineering model of moral reasoning using the idea of technological rationality in addressing a particular set of practical problems through the application of moral principles. In this approach, bioethics is a sophisticated technology to make a particular set of (potential) problems manageable and controllable. Usually the focus of ethical analysis is narrow and not too critical. For example, ethical review of research protocols is focused on informed consent, not on the social relevancy of the research; assessment of new technologies concentrates on safety, effectiveness, and costs, not on the social and ethical implications. If bioethics as a new scientific discipline and public discourse has emerged because of the development of moral problems due to the technologic advances that are changing medical care and treatment, then the outcome of this evolution is rather paradoxical (ten Have 2004). According to philosophers as Habermas, Foucault, and Illich technology confronts us with moral problems since our life-world has been penetrated, dominated, or even 'colonized' by science and technology. But when bioethics can be regarded as a specific technology itself, aimed at resolving or at least 'pacifying' the moral consequences of the use of medical technologies, it is obvious that the answer to such problems cannot be given by an ethics that is itself technologically orientated. In fact, a type of bioethics that is approaching moral problems in an engineering way, technically applying principles to cases and dilemmas, has become itself another manifestation of the same basic problem. Bioethics has become another expression of the technical rationality that has been the source of moral concerns in the first place.

C. Repositioning bioethics

This diagnosis of the development of bioethics implies that bioethics emerged as the predominant response to the criticism of contemporary medicine and the problems created by science and technology. But it also demonstrates that bioethics does not provide a cure but merely acts as a palliative. Instead of seriously addressing the philosophical queries raised within the critical movement, bioethics is primarily concerned with proposing practical answers and solutions.

The detachment from internal morality, as discussed above, was facilitated by the self-conception of bioethics as 'applied ethics'. Within the tradition of ethics this notion seems a tautology since ethics has always been considered as practical philosophy. Nonetheless, as Stephen Toulmin (1982) has argued in a famous publication, ethics has been marginalized as a sterile, academic, analytic discipline. Only through the emergence of moral problems in the medical setting it came to be revived. Emphasizing 'application' has a double connotation: it indicates that ethics is available for what we usually do, it applies to our daily problems; but ethics is also helpful, practical, in the sense that it is something to do, - it works to resolve our problems. The conception of bioethics as applied ethics not only demonstrates its usefulness (beyond mere theoretical and academic interest) but also its relevancy. It was canonized in the equally famous textbook of Beauchamp and Childress (1983) that defined biomedical ethics as the application of general ethical theories, principles and rules to specific problems which may arise in health-care delivery, research, and therapeutic practice. The aim of ethical contributions is to analyze these problems and to offer solutions that are morally justified. The main instrument of this approach is a set of moral principles. Usually three or four basic principles are used: respect for autonomy, beneficence, non-maleficence (which is sometimes included in beneficence), and justice. These principles are considered to be basic, because they are general judgments serving as justification for particular prescriptions and evaluations of human actions. Principles are normative generalizations that guide actions. From principles, ethical guidelines and rules can be derived. The advantage of the (four) principles is that they are defensible from a variety of theoretical moral perspectives. They provide an analytical framework, a universal tool, to clarify and resolve moral issues. The principles approach in analyzing moral issues is usually very helpful in identifying and mapping out the relevant moral considerations regarding medical technologies and services; it is also instructive because it points out where further studies are required. For example, in transplantation of organs from living human donors, three fundamental issues are identified: the risks and harms affecting the donor, questions about voluntary consent, and buying organs harvested from the living. The principles of beneficence and non-maleficence generate moral concerns about justifying harm to the donor. Is it justified to remove somebody's kidney when the removal harms the healthy person without producing any medical benefit to him or herself? Or is the donor more harmed by the loss of a family member or friend than by the loss of a healthy kidney? The principle of respect for autonomy generates concerns about consent. If an adult person is asked to give informed consent to surgery to remove a kidney for a family member, can the consent be truly voluntary in such circumstances? In the case of a child whose kidney is the best match for a sibling, can the parents give consent? A decision to "donate" is clearly not in the best interests of the child. Finally, the principle of justice generates concerns about the donation and transplantation systems. What criteria are used to allocate donated organs within a particular area? At the same time, it seems that commercial arrangements are increasingly used, although the sale of organs for transplantation is prohibited in many countries.

The moral issues identified by using the principles approach show that two methodological approaches need to be combined: empirical and theoretical studies. To know, for example, whether autonomy of potential donors is compromised in practice, ethicists need to engage in empirical research. To evaluate the probability and extent of harms and benefits, ethicists need to use or produce quantitative data. Insights into the factual dimensions of a technology are required before these can be assessed from normative points of view. The moral principles identify not only which facts are relevant for further consideration from a moral point view, but they also provide a normative framework for further assessment. Theoretical research here requires analysis of the philosophical and ethical literature, articulating, for example, the implications of deontologic and teleologic ethical theories with regard to the problems at hand. Usually, this is intensive and innovative work, because the existing literature has rarely foreseen or addressed the moral issues arising in present-day medicine. But apparently, even with applied ethics with emphasis on principles, there is a need to extend the ethical approach in two directions: empirical analysis of the practical context and critical philosophical understanding and justification.

D. Remedies

During the 1990s there was an intensive debate about the methodologies and approaches in bioethics. I repeatedly argued that the bioethics discourse could be richer if it explores the dialectics between internal and external morality (ten Have, 1994, 2001).

On the one hand, more attention could be paid to analyzing the internal morality of medical practices. Indeed, new approaches to bioethics have developed with focus on the particularities of such practices, such as phenomenological ethics (Zaner, 1988), narrative ethics (Newton, 1995), and care ethics (Tronto, 1993). Furthermore, traditional conceptions have been revitalized, notably the new casuistry (drawing from the classical casuistic mode of moral reasoning) (Jonsen and Toulmin, 1988), and the virtue approach, emphasizing qualities of character in both individuals and communities (Pellegrino and Thomasma, 1993). What is particularly striking is the rising interest in so-called empirical ethics. The focus of ethical research is shifting from applying ready-made ethics toward studying ethics-in-action (Arnold and Forrow, 1993). A variety of research methods is used: participatory observation, questionnaires and interviews, decision analysis, quality assessment, preference polls. The common denominator is that qualitative and quantitative data are collected via the empirical study of ethical questions. Many of these studies are fascinating since they show the underlying value pattern of specific practices and the intrinsic norms which are operative in clinical work, for example in surgery (Bosk, 1979), genetic counseling (Bosk, 1992), intensive care (Zussman, 1992), neonatal care (Anspach, 1993), and nephrology (Lelie, 1999). Especially the work of Bosk has been seminal since he introduced the methods of anthropology and sociology into bioethics (Bosk, 2008). Although empirical research in ethics can provide new and useful insights, and can be regarded as complementary to philosophical approaches, it is also troubled with fundamental problems (ten Have and Lelie, 1998). One of the basic questions concerns the moral relevancy of empirical data. Empirical research can help to explain and understand the attitudes, reasoning and motivations of the various actors in the health care setting, but empirical data in themselves can not justify how the actors ought to behave or what kind of decisions are morally justified (Pellegrino, 1995).

On the other hand, the external morality could not simply be assumed in bioethical discourse but should be critically revisited. In order to obtain a better understanding of the interaction of both moralities, internal and external, it is necessary to establish a theoretical framework relevant to medical practice in order adequately to take account of the norms and values inherent in the practice of medicine, but it requires at the same time sufficient detachment in order to provide a critical normative perspective on medical practice. This is not only true for the principles that are applied, such as the principle of respect for autonomy that is often assumed as the basic notion for ethical discourse (with individual decision-making as antidote to medical paternalism). But is seems to be true also for the emphasis on application in general. When bioethical analysis concentrates on how to morally justify the application of science and technology in the context of health care, it is often so fully immersed in the object of analysis itself that it does no longer position itself at a critical distance of scientific and technological developments. We then no longer understand how they create moral quandaries. Critical reflection on the presuppositions and implications of scientific and technological developments can clarify how moral problems emerge, why some problems emerge and not others, and how such problems are addressed. Bioethics therefore needs to go beyond the framework of science and technology itself, questioning whether the new knowledge or the specific technology, as such, is justified in the light of moral values. Here, ethical analysis does not, a priori, take science and technology for granted. It starts from a critical perspective, assuming that for example technologies are not value-neutral but incorporate particular values themselves. Technologies are expressions of values, such as the values of searching for knowledge, having offspring, or relieving suffering. However, these values are often implicitly given and not articulated. Ethical research is now taking them as the starting point for a debate on (other) motivating values in society. This type of research focuses on values underlying or embedded in the development of technology itself. For example, studies in this category will not take for granted that the progress of transplantation technologies is beneficial. They will question the specific framing of notions such as personal integrity, altruism, death, and body, which is associated with these emerging technologies. They critically examine the implied notion of 'body ownership', where the moral principle of respect for autonomy is indeed helpful to facilitate organ donation but at the same time reiterates the traditional dualistic image of the human person: an autonomous subject with a material body as its property (ten Have and Welie, 1998). These studies will also explore the recent expansion of these technologies with cell and gene transplantation. They call attention to the claims of perfectibility and immortality, often implicit in the bewildering progress of stem cell technologies, and relate such claims to a philosophical, and sometimes utopian, body of knowledge (Gordijn, 2003). The methodology of such studies is historical as well as synthetic. They attempt to provide a diachronic and synchronic perspective: values embodied in current technologies are explained in connection to similar values in history, but they are also clarified in connection to developments in other scientific disciplines, thus looking beyond the framework of present times and existing disciplines (ten Have, 1995). The presupposition of this type of ethics research is that ethics first of all is the philosophical effort to understand ourselves and our existence in terms of what is desirable or undesirable, supportable or reprehensible, good or bad.

E. From application to interpretation

The complex interactions between the internal and external morality of health care practices remind us that bioethics is first of all a philosophical activity. As a particular domain of philosophy, ethics proceeds from empirical knowledge, viz. moral experience. The moral dimension of the world is first and foremost experienced. Ethics is the interpretation and explanation of this primordial understanding. Before acting morally we must already know, at least to some extent, what is morally desirable or right. Otherwise, we would not recognize what is appealing in a moral sense. On the other hand, what we recognize in our experience is typically unclear and in need of further elucidation and interpretation.

Because of the importance of interpretation it is argued that ethics is best considered to be a hermeneutical discipline (ten Have, 1994, 2001). Ethics can be defined as the hermeneutics of moral experience. Bioethical problems in particular must be understood within the broader framework of an interpretive philosophical theory. More or less at the same time, philosophers of medicine argued that medicine itself has to be considered as a hermeneutical enterprise; it is not or not merely a natural science (Daniel, 1986; Leder, 1988; Svenaeus, 1999). The modern emphasis on information and empirical data has contributed to new understandings of diagnosis and treatment as the physician's interpretation of what concerns the patient and what can be done to help the patient. And metaphorically, the patient is conceived as a text that may be considered on different interpretive levels. Usually, the patient is understood through an anatomico-physiological model. The patient's body is made 'readable' by the use of technology. But the biomedical language of diagnosis and treatment reduces the overwhelming amount of information presented by the patient so that the standard medical case report reflects not the story of the patient's life but of the physician's relationship with the patient's illness (Poirier and Brauner, 1988).However, different interpretative models should be re-activated to do justice to the patient's experiences.

Bioethics as interpretation rather than application concentrates upon four parameters: (1) moral experience, (2) attitudes and emotions, (3) community, and (4) ambiguity. These characteristics enrich the bioethics discourse.

First of all, for ethics, the fundamental question is not so much 'What to do?' but rather 'How to live?' What is important is praxis not poiesis. The moral relevancy of our actions should not be reduced to their effects; it is also determined by an evaluation of what we do in executing our actions. For example: the issue of experimenting with human beings should not be settled by reference to future results, but should also raise the question: Why are we interested in scientific research? This change of focus implies a re-orientation from activity to passivity, from acts to attitudes and emotions. Moral experience also involves primarily feelings, for instance, of indignation, confusion or contentment; these emotional responses should be made the object of moral thinking.

Second, the interpretive reading of a patient's situation is not an individual doctor's affair. The medical prior understandings that orientate the interpretation are the sediments of traditional cultural assumptions concerning the nature of the world and the body, and the results of a specific historical evolution of medical knowledge. Interpretation presupposes a universe of understanding. This is a consequence of the so-called hermeneutic circle; in order to interpret a text's meaning the interpreter must be familiar with the vocabulary and grammar of the text and have some idea of what the text might mean. For man as a social being, understanding is always a community phenomenon: understanding in communication with others. The continuous effort to reach consensus through a dialogue with patients, colleagues and other health professionals, induces us to discover the particularities of our own prior understanding, and through that, to attain a more general level of understanding. This seems to reflect the experience of hospital ethics committees: analyzing a case in terms of moral principles leads to a stalemate but interpreting the moral experience of the concrete participants involved in this particular case usually leads to a consensus. Since the interpretation of moral experience takes place within the context of particular social practices, intimate knowledge of the historical, medical and scientific components of those practices is essential to the task of moral criticism. Ethics can not be practiced without a high degree of engagement in medical work neither without explicit attention to the social and cultural context.

Third, ethics primarily aims at interpreting and understanding moral experience. But moral experience is complex and versatile. It implies that every interpretation is tentative; it opens up a possible perspective. Definitive and comprehensive interpretation is non-existent. An interpretive approach always has an ambiguous status: more than one meaning is admitted.

Finally, interpretive bioethics will require a new rapprochement between ethics and philosophical anthropology (ten Have, 1998). From an historical perspective bioethics has emerged through various phases of philosophical criticism of modern medicine with very different manifestations: originally epistemological, then anthropological, now ethical (ten Have,1990). Particularly in health care, normative positions and moral theories are intimately connected with images of the human being. In the medical setting we cannot escape the question: what kind of human being do we want to realize in medical activities, what kind of person do we wish to respect, heal, inform, comfort in health care?

F. The need for philosophical anthropology

The image of the human person that underlies, justifies and stimulates much of everyday medicine is a universalistic and reductionist image. In this image, human beings are understood by analysing and studying anatomical structures, physiological functions, pathological aberrations, biochemical complexities or genetic locations and dislocations. The most dominant image in modern medicine is: man as mechanism. The mechanistic image of man underlying in a prototypical way clinical and curative medicine is in fact the Cartesian heritage. Considering the human body as part of material reality has been a fruitful paradigm for modern medicine. But it is important to recognise that present-day bio-ethics has emerged from criticisms of the human being as a mechanism (ten Have, 2000; 2005). It is the one-sided perspective of this image that gives rise to many bio-ethical problems. Moral issues arise from an almost exclusively technological orientation to the world and a predominant scientific conceptualisation of human life. Human beings resist the tendencies of medicine to focus primarily on their bodies and biological existence. They protest against the overwhelming power of health professionals and health care institutions, reducing patients to cases, numbers, and objects. They object to the lack of involvement of individual beings within decision-making processes, as well as to the lack of respect for individual authenticity and subjectivity. Bio-ethics has emerged as a movement to re-introduce the subject of individual patients into the health care setting, emphasizing patients' rights, respect for individual autonomy, and the need to set limits to medical power.

The paradox, however, is that we try to address the moral issues of medicine with a conception of ethics which is itself impregnated with scientific-technical rationality. The unique view of bio-ethics as 'applied ethics' or 'principlism' that has emerged during the last thirty years seems to reinforce the dominant view of human beings as mechanism, although bio-ethics itself has mainly developed from the criticisms of this image of man.

The dialectic interaction of anthropology and ethics, as emphasized particularly in the conception of interpretative ethics, may help us to regain a view of man as social being, and therefore restore the idea of moral community (Kuczewski, 1997). Our selves are constituted through the practices of the community. Cultural context and community are constitutive of the values and goals of individuals. Communal relatedness falsifies therefore the idea of the unencumbered self, the idea of self-ownership assuming that the individual as an entity exists prior to the ends which are affirmed by it. The idea that the self autonomously designs its life-project from an asocial or pre-social position, and subsequently participates in the community, is self-defeating. Without societal culture our potential for self-determination will remain empty. Ethical reflection is primarily needed to articulate the social and cultural embedded-ness of human beings and to interpret the narrative of each individual life.

3. What obligations follow from studying medicine?

Studying medicine implies in the first place being immersed in internal morality, and in particular internalizing the virtues of being a good professional. The focus on the internal morality re-iterates the view that medicine is a profession. Medical students should know that they are not just doing academic learning or scientific research, and, - as argued above -, for most of them this is precisely the reason why they have chosen to enter into this profession. In this view medicine is not a morally neutral body of knowledge and technique; its moral content cannot be derived from the general morality of society. A full account of the content of the internal morality of medicine requires further development of two constituents: the moral goals of medicine and the morally acceptable means for achieving those goals. The clinical practice of medicine is directed on a set of particular goals, a coherent range of good healing actions. As Brody and Miller (1998) have pointed out these goals should not be too narrowly identified (interpreting 'healing' as 'curing a disease'); at the same time, even a comprehensive list of goals is limiting medical activities and requiring particular moral values rather than others. Medical practice also requires internal standards of appropriate performance. Promotion of a particular goal alone is not sufficient; it should go with morally acceptable means. Brody and Miller suggest four obligations for the practice of medicine, originating in the nature of medical practice: (1) The physician must employ technical competence in practice, (2) The physician must honestly portray medical knowledge and skill to the patient and to the general public, and avoid any sort of fraud or misrepresentation, (3) The physician must avoid harming the patient in any way that is out of proportion to expected benefit, and must seek to minimize the indignity and the invasion of privacy involved in medical examination and procedures. (4) The physician must maintain fidelity to the interests of the individual patient.

Second, the profession of medicine is special since it confronts us with the human predicament in all its variety, and more often than not, its misery, pain and despair. University education should prepare students for these essential characteristics of their future work. This requires not only psychological and moral sensitivity but also critical thinking. A hermeneutical challenge will always be there in order to understand and interpret the conditions and misfortunes presented to us. But this challenge also necessitates us to take a critical distance; not to be overwhelmed by emotions or subjected to the possibilities of technological interventions but to reflect on the conditions and circumstances in which human beings are living, the positive and negative impact of science and technology on human existence, and the particularities of the social and cultural context in which problems and questions emerge.

Studying medicine therefore not only implies moral obligations (due to the internal morality of medicine) but the relevancy of external morality also demands philosophical obligations particularly as critical reflection on specific issues and analyzing them within a wider historical and human context.

4. What is the most interesting criticism?

Two interrelated critical debates concerning method and substance of ethics have been conducted, specifically within European bioethics. The first concerns the role and approach of ethics in health care, particularly as applied ethics. The 'engineering model', as described by Caplan (1983) has been advocated as the most efficient and practical approach in ethics. For example, van Willigenburg (1993) defines himself as 'ethical engineer': he has specific expertise in managing concrete moral problems. Practical ethics is focused on solving problems. It is illusory to argue that ethics should primarily be concerned with fundamental problems. Contemporary philosophy can no longer answer fundamental problems. Any distinction between fundamental and concrete problems has disappeared since concrete problems concern essential questions in real life situations asking concrete choices and decisions, not speculation and reflection. For van Willigenburg, practical ethics has emancipated from philosophy and is an effective, useful discipline of its own. Through analysis and rigorous methodology it can structure the process of deliberation and facilitate decision-making. This view of 'ethical engineering', however, seems to reduce the moral concerns of health care to problems that need to be solved. At the same time, it reduces 'living' to 'acting' as if life is one concatenation of decisions to be made and actions to be performed. In fact this view illustrates that it is not detached from philosophy but has incorporated a specific philosophy (e.g. the philosophy of techno-science) as its hallmark. The theoretical debate about methodology and conception of bioethics as applied ethics or interpretative ethics is therefore a debate among two different views of philosophy. Another critique has addressed the distinction between internal and external morality. Reinders (1993) has argued that it does not hold water. First, the values internal to medical practice are plural; they are difficult to demarcate from external ones. Second, similar values are shared with other professional relationships. Third, external values determine what good medical practice is. Rather than being criticism, Reinders' arguments illustrate the interconnections between internal and external morality. They call in fact for more research into the internal values of care practices, and this is exactly what has happened since then.

The second critical debate concerns the substance of ethics: it is the debate between liberalism and communitarianism. Is the emphasis in ethical discourse on the primacy of the autonomous individual or on community values and perspectives? In the first two decades since the birth of bioethics, apparently the liberal paradigm was dominant: each autonomous individual should determine what is valuable and good. This emphasis was understandable given the superiority of medical power and the need to create a strong counterbalance. However the emphasis became less strong due to the development of cultural sensitivity to the context of modern medicine and bioethics (Payer, 1988; Gordijn and ten Have, 2000; Stevens, 2000). Many studies showed not only that medical practices differ among countries and cultures but also fundamental notions and conceptions of disease, health and good life and death. There was growing awareness that respect for individual autonomy was recognized as fundamental principle in bioethics in western cultures but that its significance can not be assumed in other cultures. More or less at the same time, emphasis on individual autonomy became also problematic in the west. This has been an important issue in the efforts to articulate European dimensions in bioethics. In the analysis of ethical problems in care for the chronically ill or in the health care system in general, notions such as solidarity and justice traditionally play a more important role in the European than American context. But the role of the principle of respect for autonomy is also problematic in other bioethical controversies, for example the euthanasia debate or the ethics of genetic technologies and enhancement (ten Have and Welie, 2005). It makes an important difference in the ethical analysis whether the focus is on individual decision-making or on the social or communal context. Considering human beings are part of an encompassing community, - the community of all human beings (as 'cosmopolitan citizenship') or particular communities that locate us in the world (as argued by Sandel, 1996) will widen the scope of ethical discourse, often criticizing the emphasis on individual rights and liberties. For many however, this represents a conservative position since it goes against the liberalism of western societies.

5. The most important problems for future inquiry?

Bioethics currently is at a turning point. Because of the increasing internations of medical research and the processes of globalization in general, the scope and agenda of bioethics is considerably enlarged. The adoption of the Universal Declaration on Bioethics and Human Rights, unanimously adopted by UNESCO in 2005, has earmarked this new stage in the development of bioethics as a really global bioethics. In a certain sense, the original notion of bioethics initiated by Potter (1971) is revived. Many 'new' issues are now on the agenda, requiring analysis and research, such as corruption, violence, conflicts of interests, dual use, social justice, future generations, but also ecological problems such as pollution and climate change. Bioethical discourse can no longer focus on the quandaries of rich countries but has to focus on the problems of developing countries. This revival of global bioethics also underlines that bioethics no longer is solely an academic discipline but also public discourse and political concern. More reflection is needed on the transition between scientific research and decision-making in bioethics. The contemporary linkage with human rights has created new challenges and possibilities for bioethics.

The widening of scope furthermore implies a new focus on the ancient problem of universal values and principles. The adoption of the UNESCO declaration illustrates that there is now agreement about principles that form the basis of international, multicultural bioethics, itself firmly founded on international human rights, as predicted years ago by Thomasma (1997). However, the question remains whether the principles are universal as such or merely universally affirmed. It can be that they are gradually discovered to be relevant and justified everywhere, even if they have emerged in particular cultures. Or it can be the result of a gradual expansion, domination, or even imposition, of particular principles in the process of globalization. It has been argued, for example, that the UNESCO declaration gives primacy to individual interests (suggesting that it mainly reflects western perspectives). A closer look at the listed principles, however, shows that agreement was reached on a much broader range of principles, beyond the individually orientated ones. In fact, the principles can be ordered as orientated to individuals, to interaction between individuals, to society, to culture, to future generation, and to the environment. Ironically, the 'Georgetown mantra' has been complemented with social, cultural and ecological principles. It remains to be seen whether the right balance has been struck between universal human values and cultural differences, and what will be the relevancy of cultural diversity within the enlarged scope of global bioethics.

Another question that continues to return in global bioethics concerns the issue of application. Even if the search for universal principles will be successful, it remains true that from a communitarian perspective the universal human condition of existence as a communal-cultural being can only be realized in particular ways. The communitarian self is constituted by particular cultural characteristics. Even if principles are universally adopted, in practice their application must be tailored in multiple ways to accommodate different types of research and health care, categories of patients and problems, and cultural settings and traditions. Specification will be particularly important for the application of relatively new principles such as the principle of social responsibility and health. It states that progress in science and technology should advance, among other things, access to quality health care and essential medicines, access to adequate nutrition and water, and reduction of poverty and illiteracy. How such principles can be applied in heterogeneous settings and what will be their practical implications, will require new and challenging research.

At the same time, the UNESCO declaration (and other possible universal statements of bioethical principles) contains an expression of a major characteristic of bioethics: the continuous need for interpretation. Bioethical problems commonly arise because conflicts exist between several competing ethical principles. Often it is not obvious which principle will prevail. Accordingly, a careful balancing of principles is usually required. The declaration states principles that may occasionally seem inconsistent. However, ethical decision-making in practice frequently requires rational argumentation and the weighing of the competing principles at stake. In order to advance decision-making, the principles are to be understood as complementary and interrelated. This means that even if universal principles are identified, on the basis of which global bioethics may be justified, the work of bioethics has hardly begun. The challenge then is how such principles can be translated and implemented within different contexts, cultures and traditions.

References

Aghina, M.J. (1978): Patiëntenrecht. Een kwestie van gewicht. Van Gorcum, Assen.

Anspach, R.R. (1993): Deciding who lives. Fateful choices in the intensive-care nursery. University of California Press, Berkeley.

Arnold, R.M. and Forrow, L. (1993): Empirical research in medical ethics: an introduction. Theoretical Medicine 14: 195-196.

Beauchamp, T.L. and Childress, J.F. (1983): Principles of biomedical ethics (2nd ed.). Oxford University Press, New York/Oxford.

Berg, J.H. van den (1969): Medische macht en medische ethiek. Callenbach, Nijkerk.

Bleuler, Eugen (1921): Naturgeschichte der Seele und ihres Bewustwerdens. Eine Elementarpsychologie. Springer, Berlin.

Bosk, C.L. (1979): Forgive and remember. Managing medical failure. The University of Chicago Press, Chicago/London.

Bosk, C.L. (1992): All God's mistakes. Genetic counseling in a pediatric hospital. The University of Chicago Press, Chicago/London.

Bosk, C.L. (2008): What would you do? Juggling bioethics and ethnography. University of Chicago Press, Chicago.

Brody, H. and Miller, F.G. (1998): The internal morality of medicine: Explication and application to managed care. Journal of Medicine and Philosophy 23(4): 384-410.

Caplan, A.L. (1983): Can applied ethics be effective in health care and should it strive to be? Ethics 93; 311-319.

Daniel, S.L. (1986): The patient as text: A model of clinical hermeneutics. Theoretical Medicine 7: 195-210.

Dijk, P. van (1978): Naar een gezonde gezondheidszorg. Ankh-Hermes, Deventer.

Engelhardt, H.T. (1974): Explanatory models in medicine: Facts, theories, and values. Texas Reports on Biology and Medicine 32 (1): 225-239.

Gordijn, B. and ten Have, H. (eds.) (2000): Medizinethik und Kultur. Grenzen medizinischen Handelns in Deutschland und den Niederlanden. Fromman-Holzboog, Stuttgart – Bad Cannstatt.

Gordijn, B. (2003): Die medizinische Utopie. Eine Kritik aus ethischer Sicht. Nijmegen.

Have, H. ten (1980): Wijsbegeerte der geneeskunde. Algemeen Nederlands Tijdschrift voor Wijsbegeerte 72: 242-263.

Have, H. ten (1980): Self-help. Backgrounds, postulates and problems of a new phenomenon. Huisarts en Wetenschap 23: 305-308.

Have, H. ten (1983): Geneeskunde en filosofie. De invloed van Jeremy Bentham op het medisch denken en handelen. De Tijdstroom, Lochem.

Have, H. ten (1984): Ziekte als wijsgerig probleem. Wijsgerig Perspectief 25: 5-12.

Have, H. ten & van der Arend, A. (1985): Philosophy of medicine in the Netherlands. Theoretical Medicine 6: 1-42.

Have, H. ten (1990): Verleden en toekomst van medische filosofie. Scripta Medico-philosophica, Schrift 7, p. 5-18.

Ten Have, H. (1994): The hyperreality of clinical ethics: A unitary theory and hermeneutics. Theoretical Medicine 15: 113-131.

Have, H.A.M.J. ten (1995): Medical technology assessment and ethics. Ambivalent relations. Hastings Center Report, 25:13-19.

Have, H.A.M.J. ten (1998): Images of man in philosophy of medicine. In: Evans, M. (ed.): Critical reflection on medical ethics. JAI Press, Stamford (Conn.), p.173-193.

Have, H.A.M.J. ten and Lelie, A. (1998): Medical ethics research between theory and practice. Theoretical Medicine and Bioethics 19: 263-276.

Have H.A.M.J. ten and Welie J.V.M. (eds.) (1998): Ownership of the human body. Philosophical considerations on the use of the human body and its parts in healthcare. Dordrecht, Boston, London: Kluwer Academic Publishers.

Have, H. ten (2000): The zapping animal. Oscillating images of the human person in modern medicine. In: A-T.Tymieniecka and Z. Zalewski (eds.): Life - The human being between life and death. Analecta Husserliana LXIV, Kluwer Academic Publishers, Dordrecht, pp. 115-123.

Have, H.A.M.J. ten (2001): Theoretical models and approaches to ethics. In: H.A.M.J.ten Have & B. Gordijn (eds.): Bioethics in a European perspective. Kluwer Academic Publishers, Dordrecht/Boston/London, pp. 51-82.

Have, H. ten (2004): Ethical perspectives on health technology assessment. International Journal of Technology Assessment in Health Care 20(1): 71-76.

Have, H.A.M.J. ten (2005): A communitarian approach to clinical bioethics. In: C. Viafora (ed): Clinical bioethics. A search for the foundations. Springer, New York, pp. 41-51.

Have, H. ten and Welie, J. (2005): Death and medical power. An ethical analysis of Dutch euthanasia practice. Open University Press, Maidenhead (UK).

Illich, I. (1975): Medical nemesis. Calder & Boyars, London.

Jonsen, A.R. and Toulmin, S. (1988): The abuse of casuistry. University of California Press, Berkeley, CA.

Kuczewski, M.G. (1997): Fragmentation and consensus. Communitarian and casuist bioethics. Georgetown University Press, Washington, D.C.

Leder, D. (1988): The hermeneutic role of the consultation-liaison psychiatrist. Journal of Medicine and Philosophy 13: 367-378.

Lelie, A. (1999): Ethiek en nefrologie. Dissertation, University of Nijmegen.

McKeown, T. (1976): The role of medicine. Dream, mirage or nemesis? The Nuffield Provincial Hospitals Trust, London.

Newton, A.Z. (1995): Narrative ethics. Harvard University Press, Cambridge (Mass.).

Payer, L. (1988): Medicine & Culture. Varieties of treatment in the United States, England, West Germany, and France. Henry Holt and Company, New York.

Pellegrino, E.D. (1976): Philosophy of Medicine: problematic and potential. Journal of Medicine and Philosophy 1 (1): 5-31.

Pellegrino, E.D. and Thomasma, D.C. (1993): The virtues in medical practice. Oxford University Press, New York/Oxford.

Pellegrino, E.D. (1995): The limitations of empirical research in ethics. Journal of Clinical Ethics 6: 161-162.

Poirier, S. and Brauner, D.J. (1988): Ethics and the daily language of medical discourse. Hastings Center Report 18: 5-9.

Potter, Van Rensselaer (1971): Bioethics. Bridge to the future. Prentice-Hall, Englewood Cliffs, New Jersey.

Reinders, J.S. (1993): Over de medische praktijk als uitgangspunt van ethische reflectie. In: F.W.A. Brom, B.J. van den Bergh & A.K.Huibers (eds.): Beleid en ethiek. Van Gorcum, Assen, pp.226-232.

Sandel, M.J. (1996): Democracy's discontent. America in search of a public philosophy. The Belknap Press of Harvard University Press, Cambridge (Mass.) and London.

Sporken, P. (1969): Voorlopige diagnose. Inleiding tot een medische ethiek. Ambo, Bilthoven.

Stevens, M.L.T. (2000): Bioethics in America. Origins and cultural politics. The Johns Hopkins University Press, Baltimore and London.

Svenaeus, F. (1999): The hermeneutics of medicine and the phenomenology of health. Steps towards a philosophy of medical practice. Linköping University, Linköping (Sweden).

Thomasma, D. (1997): Bioethics and International Human Rights. Journal of Law, Medicine & Ethics 25: 295–306.

Toulmin, S. (1982): How medicine saved the life of ethics. Perspectives in Biology and Medicine 25: 736-750.

Tronto, J.C. (1993): Moral boundaries. A political argument for an ethic of care. Routledge, New York/ London.

Verbrugh, H.S. (1978): Paradigma's en begripsontwikkeling in de ziekteleer. De Toorts, Haarlem.

Willigenburg, T. van (1993): Ik ben een ethisch ingenieur! In: F.W.A. Brom, B.J. van den Bergh & A.K.Huibers (eds.): Beleid en ethiek. Van Gorcum, Assen, pp.189-204.

Wulff, H.R. (1980): Principes van klinisch denken en handelen. Bohn, Scheltema & Holkema, Utrecht.

Zaner, R.M. (1988): Ethics and the clinical encounter. Prentice-Hall, Englewood Cliffs, N.J.

Zola, I.K. (1973): De medische macht. Boom, Meppel.

Zussman, R. (1992): Intensive care. Medical ethics and the medical profession. The University of Chicago Press, Chicago/London.

7
Bjørn Hofmann

Professor

Center for Medical Ethics, University of Oslo and Department of Health, Technology, and Society, University College of Gjøvik, Norway

1. **Why were you initially drawn to Philosophy of Medicine?**

When I asked a pathologist at a county hospital in Norway where I was working during the 1990s "what is disease?" he handed over the International Classification of Disease (ICD-9), and when I asked the physiotherapist "what is health?" she started telling me about new techniques in recent rehabilitation programs. As I queried the gastroenterologist who cared for his gastroscopes like little babies, he claimed that they were value-neutral means to an external end. This surprisingly vague comprehension of key concepts in health care and the naïve conception of health technology spurred my interest in philosophy of medicine and lead me back to the interests of my adolescence: philosophy and sociology, interests which I had abandoned in order to lead an independent and responsible life.

As many youngsters (growing up during the 1970s) I wanted to be independent in spirit and earnings. This did not seem possible studying the heavily disciplined culture of the humanities[1], whereas an education in the natural sciences seemed to be a good way to facilitate an independent and self-supportive life. The natural sciences and technology were under heavy criticism at this time, and I was firmly determined to enter a field where science and technology was used for good and not for bad. Medicine seemed to be such a field, and I focused my studies and later my work on medical technology. Of course I soon discovered that medical technology is not only for good. At the same time I realized that many of my otherwise diligent and reflective colleagues in the

Norwegian health care system did not have any reasonable grasp of key concepts of their professions (e.g. health, disease, effectiveness, utility). To me health care practice appeared to be happily isolated from any influence of the conceptual analysis common in basic philosophy and from the ample insights on the blurred relationship between facts and values in the science and technology studies. Therefore, the vague and undefined concept of disease[2-5] and the surprisingly strict differentiation between facts and values in the practical implementation and use of health technology revitalized my interest in philosophy. In the literature which by now has become a canon for philosophy of medicine, I found encouragement and comfort and a series of fascinating frameworks for addressing the issues I felt pressing but ignored in health care practice. Hence, my interest in disease[6], technology[7] and values was what initially drew me to Philosophy of medicine[8]. Later I have had the pleasure of working with a series of other topics both in philosophy of medicine and in bio(medical) ethics, e.g. conceptual analysis of autonomy[9-11], health care needs[12], uncertainty[13-14] and futility[15-16] as well as assisted reproductive technologies (ART), informed consent[17-19] and organ donation[20].

2. What does your work reveal about Philosophy of Medicine that other academics, citizens typically fail to appreciate?

Philosophy of medicine is not a clearly defined field with a crisp demarcation. It covers a vast variety of topics, theories and perspectives. Only few scholars cover it all. My work is limited to some selected areas in the interesting intersection between epistemology and ethics. I have tried to uncover how technology, methodology, heuristics, research design, and basic concepts which health care professionals tend to think are descriptive and value neutral are prescriptive and value-laden. This is done in areas as different as intracytoplasmatic sperm injection (ICSI)[21], proton therapy[22], positron-emision-tomography (PET)[23], palliative[24] and bariatric surgery[25-26], ultrasound screening[27-28], new-born screening[29], HPV-vaccination[30], and preimplantation genetic diagnosis (PGD)[31] to mention a few. I have tried to show how what characterizes science and technology is not only its scientific methodology and technical specifications (i.e. the professional norms), but also its moral norms[32]. Reasoning in science and technology does not only implicate scientific inference, but also analogical expansion[33-35]. Analysing the fact-value-distinction is commonplace in philosophy of science and in science and technology studies, but their insights are not acknowledged in medicine and health care[36]. Therefore I think it is important to communicate these insights in a manner that appears and relevant understandable to health care professionals. As most health care professionals are preoccupied with methods, and research designs heuristics, I have found it fruitful to reveal the evaluative assumptions of their methodological choices and key concepts[37-40]. The aim is to expose the interplay between professional and moral norms in order to increase awareness, openness, and transparency.

At the same time as key insights from philosophy can be valuable for health care practice, the challenges in the practically oriented field of health care can be of great value to challenge and refine basic conceptions in philosophy. Hence, I think that philosophy of medicine has a twofold task, i.e. both to communicate and convey insights from philosophy into health care, and to let the challenges of health care practice originate and refine philosophical.

One area where I have found it particularly important to acknowledge and address this epistemic-ethical duality of health care science and technology is in health technology assessment (HTA)[41-44]. HTA is itself oftentimes handled as a value-neutral means in order to provide an independent value-laden end (i.e. value neutral information about the effectiveness, efficiency, and safety of health technologies). This makes the field prone to hidden values and agendas. To uncover these and facilitate a transparent process of technology assessment, appraisal and implementation has been an important task. At the same time single theories or particular approaches in philosophy appear insufficient to do the job. Standard theories tend to focus on one (or few) principles and restricted perspectives. The challenges in health care are complex as are its technologies. Hence, single principles and narrow perspectives are prone to ignore important issues. Multiple perspectives and multitude of theories can therefore be of great value[45-46].

In particular I have implemented the insights from my research on technology and values in a method for addressing ethical issues in HTA (a Socratic, axiological method), which aims at highlighting moral aspects of health technologies and its assessment and implementation.[41-42]

Another field where the interest in the epistemic-ethic relationship has spurred empirical investigations is in radiological imaging. There is a substantial variation in the use of radiological services in Norway and in the world which cannot be explained by variations in morbidity[47-48]. Hence, there is ample room for overuse and underuse which has inspired empirical investigations of the causes[49] and reasons[50] for increased and unnecessary use of radiological services, as well as conceptial analyses of diagnostic futility[15]. These studies show that technological development and access to imaging services are key factors for the use and overuse of advanced health services. Radiology has turned out to be a brilliant field to study the interplay between facts and values in the health care setting.

3. What, if any, practical and/or social-political obligations follow from studying medicine from a philosophical point of view?

As indicated, studying medicine from a philosophical point of view can reveal implicit assumptions, hidden agendas, and concealed premises in health care. Such academic disclosures may oblige to take part in public debates on health care (technology) in order to ensure an open and transparent debate. Industry, professionals, patient interest groups, health authorities, and health policy makers all have strong interests and are partly good at promoting them. When hidden agendas and one-sided arguments dominate public debates on health care, it may become imperative for academics to take part in public debates to highlight the implicit premises and display the neglected arguments. This may require leaving the cosy and safe armchair and move on to unfamiliar ground in public debates as well as getting quite a few hard beats.

Secondly, it is important to communicate the insights from philosophy to the health care field. However, this has to be done in a way that is comprehensible to the health care setting. It is too easy to claim that health care professionals are ignorant of commonplace insights in philosophy. We ought to communicate in a way that is intelligible and relevant for the field. Philosophy of medicine is no armchair philosophy. It is a bedside, lab-bench, policy-making, and blood splatting philosophy. We need to speak to the patients, the health professionals, and the health policy makers, but not to run their errand.

Thirdly, philosophy of medicine has to challenge medicine and health care on areas where practice lacks justification or counters basic insights, principles, and perspectives. This appears relevant in fields as different as human enhancement, stem cell research, and euthanasia, as well as in the overuse of diagnostic tests. In practice it is easy to find cases where suffering individual need help from new and emerging technologies, but where it's implications are incomprehensible and may violate basic individual rights as well as social values.

Fourthly, philosophy has to let itself be inspired and challenged by the conceptions and challenges in health care practice. When theories, positions or perspectives are inadequate to address practical challenges, they should be replaced or revised. Health care can save the life of philosophy in the same way that medicine saved the life of ethics[51].

Fifthly, some of the insights in philosophy of medicine have practical and political implications. Phenomenological insights in patient perspectives on illness can guide care, treatment, and policy making. Empirical and philosophical insights into consciousness and volition may have implications for conceptions of responsibility for actions and health.

4. What do you see as the most interesting criticism against your own position in philosophy of medicine?

There are many relevant and interesting criticisms against a pluralistic or eclectic approach. If you accept more than one perspective, position or theory to be relevant, as I do, you will meet the critique of these positions, be confronted with the task of balancing these perspectives, positions or theories, and you will be charged with being a relativist. Adhering to one theory or tradition appears more lucrative. However, this may make us miss important aspects of the complexities in modern health care (and its complex technologies). Moreover, single theories are prone to a wide range of challenges with specification and application, running into the same problems with interpretation and balancing as pluralist approaches. Hence, these are challenges that we all share. Furthermore, the selection of particular perspectives poses problems which a monothetic approach avoids. To argue that the selection and use of particular perspectives is based on the subject matter, in an Aristotelian sense, does not solve the problem, as we still have to justify *how* the subject matter directs the selection of theories, perspectives, and positions. On the other hand: "to the man with the hammer, everything becomes a nail." A "triangulation" of various approaches and perspectives may reveal important aspects that monothetic approaches ignore.

This illustrates that my approach in philosophy of medicine is open to a wide range of criticism. However, this is not different from many other approaches. The One-Great-Answer to all challenges in medicine, health care, and the life sciences has yet to be found.

Accordingly, displaying and revealing underlying epistemic-ethical structures and premises is by no means value-neutral. Accordingly, the uncovering process can itself have unrevealed premises. Hence, the premises and assumptions of the process itself have to be uncovered. That is, the value-ladenness of a Socratic approach has to be spelled out.

Additionally, it can be argued that any valid approach in philosophy of medicine should be normative, and that a norm-revealing approach is too descriptive and weak. However, any normative approach should have an explicit justification[52], and whenever any position in philosophy of medicine claims to have valid perspectives and preferable solutions in health care they need to be warranted, and it is not obvious that philosophy of medicine qua subject or perspective has a more proficient voice than other subjects or perspectives. If philosophers have prominent voices in health care issues, this needs justification.

5. With respect to present and future inquiry, how can the most important problems concerning Philosophy of Medicine be identified and explored?

Philosophy of medicine has become extensive in topics, themes and methods and there will be many ways that important problems will emerge. New technologies, such as biotechnology manipulating basic characteristics of cells, will enhance existing challenges and pose new ones. iPSC, chimeras, and cybrids are but some examples. Such technologies challenge existing entities and conceptions, e.g. of what biological material is, and urges us to find new ones, e.g. biological rights[53]. The application of traditional technologies in new fields, such as the use of fMRI to study consciousness, emotions, and volition will identify new problems. This field has already sparked interesting debates on free will and agency.

Moreover, combination of perspectives in the philosophy of medicine may also come to identify new areas of research. Importantly, pressing problems in health care may present the most challenging and relevant problems to philosophy of medicine. New and emerging phenomena in clinical practice, such as apathetic children among Swedish asylum seekers, may be but one example.

Yet another way that problems for philosophy of medicine may occur is through paradoxes in public debates on health, in professional disputes, in health policy making, as well as in philosophical arguments.[54] Contradictions and opposing principles are issues for philosophy of medicine. Hence, there are many ways that problems concerning philosophy of medicine may be identified and explored, and we should refine our sensibility and develop our methods to address the important problems to come.

References

[1] Snow CP. The two cultures, 1959.

[2] Hofmann B. The concept of disease-vague, complex, or just indefinable?. Medicine, Health care and Philosophy 2010;13(1):3-10.

[3] Hofmann B. On the triad disease, illness and sickness. Journal of Medicine and Philosophy 2002; 27(6): 651-74.

[4] Hofmann B. Complexity of the concept of disease as shown through rival theoretical frameworks. Theoretical Medicine and Bioethics 2001; 22(3): 211-37.

[5] Hofmann B. Simplified models of the relationship between health and disease. Theoretical Medicine and Bioethics 2005; 26(5): 355 - 377.

[6] Hofmann B. Hva er sykdom? [What is disease?] Gyldendal Akademisk, 2008.

[7] Hofmann B. Infusjonsapparatur. Oslo: Universitetsforlaget, 1998. [Infusion Devices. Oslo: Scandinavian University Press, 1998].

[8] Hofmann B. The technological invention of disease - on disease, technology and values. Thesis. Oslo: University of Oslo. Feb 28 2002.

[9] Hofmann, Bjørn. Det moralske grunnlaget for å vurdere samtykkekompetanse. [The moral basis for capacity to consent assessments] Nordic Journal of Applied Ethics / Etikk i praksis 2007;1(1):33-48.

[10] Pedersen, Reidar; Hofmann, Bjørn; Mangset, Margrethe. Pasientautonomi og informert samtykke i klinisk arbeid. [Patient autonomy and informed consent] Tidsskrift for Den norske lægeforening 2007;127(12):1644-1647.

[11] Hofmann B, Lysdahl KB. Moral principles and medical practice: the role of patient autonomy in the extensive use of radiological services. Journal of Medical Ethics 2008; 39: 446-449.

[12] Frich JC, Hofmann B. [Needs – disease without limits] Behov - sykdom uten grenser? I: Penger og verdier i helsetjenesten. Gyldendal Akademisk 2009 ISBN 978-82-05-39332-5. s. 54-67.

[13] Hofmann B. For sikkerhets skyld – om skylden i vår søken etter sikkerhet. ["Just in case": – about the challenges in our strive for certainty] Bibliotek for Læger 2005; 197(4). 353-64.

[14] Hofmann B. Sikkerhetens skyld: Om handling under usikkerhet. [How safe is "better safe than sorry"?] Utposten 2010; 1: 32-37.

[15] Hofmann B. Too much of a good thing is wonderful? A conceptual analysis of excessive examinations and diagnostic futility in diagnostic radiology. Medicine, Health care and Philosophy 2010;13(2): 139-148.

[16] Waaler D, Hofmann B. Image Rejects – Radiographic Challenges. Radiation Protection Dosimetry 2010; 1: 1-5.

[17] Hofmann B. Broadening consent and diluting ethics. Journal of Medical Ethics 2009; 35(2): 125-129.

[18] Hofmann B. Bypassing consent for research on biological material. Nature Biotechnology 2008; 26:979-980.

[19] Hofmann B, Solbakk JH, Holm S. Consent to Biobank Research: One Size Fits All? I: The Ethics of Research Biobanking. Dordrecht Heidelberg London New York: Springer 2009 ISBN 978-0-387-93871-4. s. 9-29

[20] Hofmann, Bjørn. The moral challenges with living kidney donors. Tidsskrift for Den norske lægeforening 2007;127(22):2964-2966.

[21] Hofmann B. Technology assessment of intracytoplasmic sperm injection - an analysis of the value context. Fertil Steril. 2003 Oct;80(4):930-5.

[22] Hofmann B. Fallacies in the arguments for new technology: the case of proton therapy. Journal of Medical Ethics 2009; 35: 684-687.

[23] Hofmann B. PETitio principii. Dagens Medisin 2002; 10 (6.juni): 29.

[24] Hofmann B. Håheim LL, Søreide JA. Ethics of palliative surgery in patients with cancer. British Journal of Surgery 2005; 92(7): 802-9.

[25] Hofmann B. Stuck in the Middle: The Many Moral Challenges With Bariatric Surgery. American Journal of Bioethics 2010;10 (12):3-11.

[26] Hofmann B. The Encompassing Ethics of Bariatric Surgery. American Journal of Bioethics 2010; 10 (12):W1-W2.

[27] Hofmann B. Descriptive and normative fallacies in presenting detection rates. Ultrasound in Obstetrics and Gynecology 2009; 34(2):237-237.

[28] Reinar LM, Smedslund G, Fretheim A, Hofmann B, Thürmer H. Rutinemessig ultralydundersøkelse i svangerskapet. [Ultrasound sceening during pregnancy] Oslo: Nasjonalt kunnskapssenter for helsetjenesten 2008 (ISBN 978-82-8121-203-9) 129 pages. Rapport fra Kunnskapssenteret. [Norwegian Knowledge Centre for the Health Services, in Norwegian with an English summary]

[29] Hofmann B. Nyfødtscreening – mer skjult tvang? [Newborn Screening: More covert enforcement?] Tidsskrift for Den norske legeforening 2010; 130(3): 291-293.

[30] Hofmann B. Vaksiner mot humant papillomavirus (HPV). Etiske aspekter ved innføring av profylaktiske HPVvaksiner. [Ethical aspects of HPV vaccination]. Oslo: Nasjonalt kunnskapssenter for helsetjenesten 2008 (ISBN 978-82-8121-217-6) 29 pages. [Norwegian Knowledge Centre for the Health Services]

[31] Hofmann B. The paradoxes of PGD. Submitted manuscript.

[32] Hofmann B. That's not science! The role of moral philosophy in the science/non-science divide. Theor Med Bioeth. 2007;28(3): 243-56.

[33] Hofmann B, Holm S, Solbakk JH. Analogy is Like Air—Invisible and Indispensable: Response to Open Peer Commentaries on "Analogical Reasoning in Handling Emerging Technologies: The Case of Umbilical Cord Blood Biobanking". American Journal of Bioethics 2006;6(6):W13-W14.

[34] Hofmann B, Solbakk JH, Holm S. Analogical reasoning in handling emerging technologies: the case of umbilical cord blood biobanking. American Journal of Bioethics 2006;6(6):49-57.

[35] Hofmann B, Solbakk JH, Holm S. Teaching old dogs new tricks: The role of analogies in bioethical analysis and argumentation concerning new technologies. Theoretical Medicine and Bioethics 2006;27(5):397-413.

[36] Hofmann B. Is there a technological imperative in health care? International Journal of Technology Assessment in Health Care 2002; 18(3): 675-89.

[37] Hofmann B. The inference from a single case: moral versus scientific inferences in implementing new biotechnologies. Medical Humanities 2008 ; 34: 19-24.

[38] Hofmann B. When means become ends: technology producing values. Seminar.net - Media, technology and lifelong learning 2006;2(2).

[39] Hofmann B. On the value-ladenness of technology in medicine. European Journal of Medicine, Health Care and Philosophy 2001; 4(3): 335-345.

[40] Hofmann B. Technological medicine and the autonomy of man. European Journal of Medicine, Health Care and Philosophy. 2002; 5: 157-67.

[41] Hofmann B. Simplified models of the relationship between health and disease. Theoretical Medicine and Bioethics 2005; 26(5): 355 - 377.

[42] Hofmann B. Toward a procedure for integrating moral issues in health technology assessment. International Journal of Health Technology Assessment 2005;21(3):312-18.

[43] Hofmann B. Etikk i vurdering av helsetiltak. Utvikling av en metode for å synliggjøre etiske utfordringer ved vurdering av helsetiltak. [Ethics in Health Technology Assessments (HTA)] Oslo: Nasjonalt kunnskapssenter for helsetjenesten [Norwegian Knowledge Centre for the Health Services] 2008 53 pages.

[44] Anttila, Heidi; Autti-Rämö, Ilona; Kristensen, Finn Børlum; Cleemput, Irina; Laet, Chris de; Hofmann, Bjørn. HTA Core Model for Medical and Surgical Interventions. Danmark: European network for Health Technology Assessment 2007. 365 pages.

[45] Hofmann B. Medicine as téchnê - a perspective from antiquity. Journal of Medicine and Philosophy 2003; 28(4):403-25.

[46] Hofmann B. Medicine as practical wisdom (phronesis). Poiesis & Praxis 2002; 2, 135-149.

[47] Børretzen I, Lysdahl KB, Olerud HM. Radiologi i Noreg: undersøkingsfrekvens per 2002, tidstrendar, geografisk variasjon og befolkningsdose. StrålevernRapport 2006:6. [Radiology in Norway - examination frequency per 2002, trends in time, geographical variation and population dose. - In Norwegian]. Østerås: The Norwegian Radiation Protection Authority.

[48] Almén A, Friberg EG, Widmark A et al. Radiologiske undersøkelser i Norge per 2008. Trender i undersøkelsesfrekvens og stråledose til befolkningen. StrålevernRapport 2010:12. [Radiology in Norway anno 2008. Trends in examination frequency and collective effective dose to the population. - In Norwegian]. Østerås: The Norwegian Radiation Protection Authority.

[49] Lysdahl KB, Hofmann B. What causes increasing and unnecessary use of radiological investigations? a survey of radiologists' perceptions. BMC Health Services Research 2009; 9: 155. http://www.ncbi.nlm.nih.gov/pmc/articles/PMC2749824/pdf/1472-6963-9-155.pdf

[50] Lysdahl KB, Hofmann B, Espeland A. Radiologists' responses to inadequate referrals. European Radiology 2010; 20(5): 1227-1233

[51] Hofmann B. Technological paternalism: On how medicine has reformed ethics and how technology can refine moral theory. Science and Engineering Ethics. 2003; 9(3): 343-352.

[52] Nagel, T. Moral Conflict and Political Legitimacy. Philosophy and Public Affairs 1987; 16 (3), 215–40.

[53] Solbakk JH, Holm S, Hofmann B. The Ethics of Research Biobanking. Dordrecht Heidelberg London New York: Springer 2009 (ISBN 978-0-387-93871-4) 363 pages.

[54] Hofmann B. The paradox of health care. Health Care Analysis 2001; 9: 369-86.

8

Søren Holm

Professor

Centre for Social Ethics and Policy, School of Law, University of Manchester & Center for Medical Ethics, University of Oslo

---

1. Why were you initially drawn to Philosophy of Medicine?

As a child I was an avid reader across the whole range of reading material from comics and crime novels to more serious literature and from the practice of angling to quantum mechanics (for beginners), but I don't think I was particularly philosophically minded. I was undoubtedly clever, inquiring and interested in knowledge and understanding but not in any philosophically sophisticated way.

I grew up in a Pentecostal church and although there was plenty of room for discussion of important issues concerning values and ethics it was also an environment in which too radical questioning was difficult. Pentecostals were, at that time fairly sceptical towards higher education which was seen as a danger to true belief. At home, however there was always openness to discuss the great issues of the day. Both my parents had wanted to pursue academic careers but had not had the economic opportunity to do so but they were both well read and interested in politics and societal issues. My mother's political interests later led her into the research ethics committee (REC) system in Denmark where she became vice-chair of the REC for Copenhagen and a member of the national Central REC.

During my years in high school I developed my talent for irony and clever comment and I realised that whatever beliefs you have they have to be justified in the public arena of ideas.

However, until a couple of years into my medical studies my future career plan was completely fixed and it did not include philosophy in any shape or form. The only thing I ever wanted to become as a child and adolescent was a medical doctor and preferably an orthopaedic surgeon. I am born with severe bilateral clubfoot and had a number of operations as a child and my mother is a nurse so I had a strong affinity to orthopaedics. After starting my medical studies I quickly realised that orthopaedic surgery was not for me. The old joke that "an orthopaedic surgeon has to be strong as an ox.... and almost as bright" is not true, but never the less I could see that orthopaedic surgery would not offer me much in the way of intellectual stimulus. So I set my sights on some sub-specialty in internal medicine or perhaps neurology. That I became interested in philosophy of medicine and began to read and think about philosophy more seriously was due to three fairly independent things that happened while I was studying medicine.

The first of these was that I and my reading partner Steen H. Hansen (now working in cell biology at Harvard) decided to follow the then optional course in medical philosophy at the University of Copenhagen. At that time it was taught by the "dream team" of Henrik R. Wulff, Stig Andur Pedersen and Henning B. Andersen and it was great! Because it was optional there were only very few people attending and we were a highly self-selected group. This made the discussions very interesting. The textbook for the course at that time had a clear Marxist inspiration so there was plenty to discuss since the popular pendulum was swinging rapidly away from Marxism and towards much more conservative ideas. It may seem incredible now, but when I started medical school in Copenhagen in February 1982 the Communist Students, the student wing of the Moscow oriented Danish Communist Party still held induction weekends for new medical students. However, attending the course on medical philosophy inspired me to start reading philosophy books in my spare time.

The second thing that happened was that I was fortuitously recruited to a research group studying the anatomy and function of the prefrontal cortex. This group was led by Ivan Divac and Jesper Mogensen and was an extremely exciting place to work. Ivan Divac marked my physiology exam paper and thought that the answer I had given to a question in neurophysiology showed some promise so he sent me a letter asking whether I was interested in doing research. I contacted him and started working in his group on the anatomy and function of the prefrontal cortex in the rat and I would probably still be working in experimental neurophysiology today if I had not become allergic to rats. Waking up in the middle of the night not being able to breathe easily was the event that eventually convinced me to seek a different career path. The kind of functional neurophysiology which the group worked with raises numerous methodological and philosophical issues. What do we mean when we say that something is an animal model of a human condition? Does it make sense to ascribe functions to parts of the brain, and if so how do you define those parts? What are the ethical issues raised by the use of animals in research and how are they intertwined with methodological choices? And of course the biggest of them all, what is the relation between brain function and mental phenomena? I discussed all of these issues and many more with Jesper Mogensen who was and is a true polymath and who taught me a lot about the necessity of being willing to question commonly held scientific assumptions and the necessity of hard work if you want to have a career in academia.

The third thing that happened was that the University of Copenhagen published a prize question on "The ethical issues in prenatal diagnosis" in 1985 as part of the long list of prize questions for that year. Because I come from a totally un-academic background I did not realise that many/most prize questions are put forward with a specific person in mind, so I decided to answer the question. My reading partner Steen Hansen was interested in one of the other questions so we both took 6 months out from our studies to do the research and write the submissions. During this process I found that medical ethics was even more interesting and complicated than I had believed before, and getting a silver medal for the submission somewhat reinforced my belief that I had some talent in the area, and some of the argument formed the basis for my first paper (1). No one else got a prize for answering that specific question, but I later found out who the question had actually been written for and also that he had actually submitted an answer. I should probably mention that my parents were rather proud that their son got to shake the hand of the Queen as part of the prize giving ceremony.

Why did I stick with philosophy of medicine instead of pursuing what would undoubtedly have been a more lucrative clinical career? This is a question I often ask myself at the end of the month when the mortgage has to be paid, but I think the fact that I have always worked together with nice people and that most people in the field are friendly and nice is a major reason. At the beginning of my career I had the great fortune to work with and alongside Povl Riis, Henrik R. Wulff, Peter Rossel and John Harris all of whom, although very eminent were also very open minded and willing to engage with a young medical doctor who, at that point believed that he knew more about philosophy than he actually did. I was also lucky enough to get the fellowships, grants and later permanent positions that allowed me to follow my interests.

2. What does your work reveal about Philosophy of Medicine that other academics, citizens, or economists typically fail to appreciate?

For me philosophy of medicine is the overarching category encompassing all forms of philosophical analysis of phenomena in medicine and health care, thus including medical epistemology, philosophical aspects of medical decision making, medical ethics, and the philosophical analysis of medical concepts. In this endeavour both analytic and continental approaches are often very valuable.

Health care professionals usually underestimate the degree to which their knowledge is uncertain and their concepts are fuzzy and ill defined and they usually overestimate the degree to which their conceptions of illness, disease and symptoms are the same as those held by the general public. But if we take a seemingly banal example such as "fever" it is obvious that it is likely to be embedded in two quite different semantic webs in "lay" and professional discourse; and that the question "Does Jim have fever?" may therefore mean something very different to the professional who asks and the person who answers.

A similar problem often occurs in writing about medical ethics where some ethicists with a background almost exclusively in moral philosophy often writes as if medical concepts have fixed and straight forward meanings, or as if medical acts admit of definite description. On the other hand health care professionals often want far more definite answers than philosophy is likely to give.

The vagueness of our terms and their mutually overlapping meanings has quite wide ranging implications for clinical decision making, for the proper understanding of informed consent and for the practice and ethics of health promotion.

One of the main tasks of philosophy of medicine is to point to the problems that arise if the complexities of medical semantics are not taken seriously and I hope that my work has contributed a little to this task.

In my recent work I have also become interested in the topic of path dependence in relation to discussions about 1) the role of medicine and its potential normative implications (i.e. is there an internal normativity in medicine?) and 2) resource allocation in medicine. In discussions of these issues we often forget that we are where we are because of a long historical process that has created the current status quo and more importantly that where we are to some extent determines were we can go in the future. I think there is a tendency to underestimate how much our current choices are actually constrained by what has gone before.

3. What, if any, practical and/or social-political obligations follow from studying medicine from a philosophical point of view?

I think that there are two very different ways in which you can approach philosophy of medicine:

1. You can use medicine primarily as an area in which interesting philosophical issues arise that can be used to hone and improve philosophical analysis or where you can find interesting examples that illustrate controversies in philosophy, or

2. You can engage in philosophy of medicine in order to illuminate issues that are of practical importance in health care practice.

Both types of approach are valid and can lead to interesting insights but I have always been much more attracted to the second approach, partly for the reason that Karl Marx stated very clearly "The philosophers have only interpreted the world, the point, however, is to change it.", and partly I think because I came to the field from a medical background with an eye to the many problems that doctors and other health care professionals face on a daily basis.

I don't think that any particular practical obligations follow from the first approach to philosophy of medicine, apart from the very general scholarly obligations 1) not to claim any more practical importance for your results than they actually have and 2) not to propose that policy should be based on your results unless you fully understand the complex web of social activities that constitute medicine and health care in practice. To take an example, it may well be that the concept of mental illness is philosophically puzzling and perhaps in the final analysis incoherent. But very little follows from this for how we should relate to and as a society take care of those people who, for instance experience the complex of symptoms that psychiatrists call schizophrenia. They may not have a mental illness as per our analysis, but they do suffer and if we can deal effectively and ethically with their suffering it matters relatively little that we may have misclassified its cause.

The second approach to philosophy of medicine does, in my view come with a set of obligations or values attached. If your stated aim is to illuminate issues that are of practical importance in health care practice it seems to follow that there is some commitment to work on issues that are important, either because they cut across many areas of medicine (e.g. the concepts of health and disease, aspects of clinical decision making, informed consent, resource allocation etc.), or because they are crucial to practice in one particular area (e.g. abortion).

A second obligation that flows from this approach is an obligation to take the practicalities of the health care setting seriously in both analysis and conclusions. Health care systems are hierarchical and the personnel often work under severe time constraints and it is of little use to health care professionals if philosophers analyse the issues in the abstract without taking account of these features, or if they just gesture towards the possibility that more time and money could be found. Danish social-democrat politicians of a former generation often referred to the need to "understand the real conditions in the steel industry" before proposing policies on labour related issues. Similarly medical philosophers who claim to do work that is directly relevant to health care professionals and patients undoubtedly have to understand and take account of the real conditions in the health care industry (and since I am not a postmodernist or constructivist I do not have to worry too much about to what degree the real conditions are really real or constructed).

I also, slightly more contentiously believe that there is an obligation to take issues of justice in health care seriously in work within the philosophy of medicine. The patient's need for health care and the often very significant negative welfare effects of illness make claims for health care strong claims. But justice issues are notoriously difficult and there is a tendency to bracket them and proceed with the analysis of, for instance cognitive enhancement as if concerns about justice can wait or can be sorted after we have decided whether cognitive enhancement is worth having or not. But if history teaches us anything it is that justice is only ever achieved if injustice is taken seriously as a problem that must be addressed.

Finally I think that work in philosophy of medicine has to be mindful of the often profound difference between the first and the third person perspective in relation to medical matters. It is important to understand the lived world of patients and health care professionals and not to replace it with the philosopher's "view from nowhere".

4. What do you see as the most interesting criticism against your own position in philosophy of medicine?

The most interesting criticism against my position in the philosophy of medicine is that it is not sufficiently clear what that position is! I think this is a valid, but at the same time slightly misguided critique. It is true that I have written about a fairly wide range of topics, from the implications of chaos theory for the philosophy of medicine to the relevance of K.E. Løgstrup for nursing philosophy (2,3). And it is true that I have never nailed my flag to any particular theoretical mast (although I have argued for definite conclusions), but I still think that there is a common underlying approach in my research and writing.

I would like to think of myself as the "thinking man's conservative", my position as "defensible conservatism" (conservatism here used without any political connotations!) and my approach as "rigorous scepticism towards sweeping claims". What I mean by this is that quite radical and wide ranging claims are often made in philosophy of medicine along the lines of "chaos theory will revolutionise our understanding of health and disease", "a proper understanding of moral status will show that foetuses have no moral standing", "true informed consent is impossible" or "the theory of philosopher X is the (only) right theory to use in understanding problems of type Y". Many such claims come from attempts to find one master argument, a "hole in one" argument that will solve a whole range of problems. But most of such claims are either 1) implausible or 2) have obvious implications that even their most ardent proponents want to deny (or hide) or 3) have far more restricted scope than is claimed.

A lot of my research and writing has been aimed at showing that specific radical claims suffer from one or more of these problems. In many cases this is a fairly a-theoretical exercise. You do, for instance not necessarily have to commit to any particular analysis of the concept of disease to show that chaos theory is unlikely to revolutionise our understanding of that concept. It is enough to show that the kind of new knowledge generated by chaos theory can be easily accommodated within many (or all) of the standard analyses of the disease concept.

One common result of such a sceptical and deflationary analysis of sweeping claims is that although the most radical version of the claim can be shown to be problematic, or sometimes downright false a more restricted claim is more plausible. I have often in my critical writings tried to point out what such a more restricted claim could be and how it could fit within a defensible conservatism.

Now and again I have felt the pull of grand theory, of the all encompassing, irrefutable and utterly compelling "Holm approach to philosophy of medicine". But the temptation usually passes fairly quickly. If we restrict the scope to just medical ethics there is the immediate problem that all of the available grand theories in the field, consequentialism, libertarianism, Kantianism, Gewirthianism etc. have significant justificatory problems and all generate immensely counterintuitive results. Results that are so counterintuitive that the theories have to be patched up with various ad hoc patches to provide moral guidance that is in any way appropriate to normal life and normal medical practice.

5. How can the most important problems concerning Philosophy of Medicine be identified and explored?

In philosophy of medicine we often run up against what John Woods has called "The most difficult problem in philosophy", that is the problem of deciding how to assess a set of arguments that lead to a truly surprising, radical and potentially disturbing conclusion (4). Should we "follow the argument where it leads" and accept the conclusion or should we see conclusion as an instance of a reductio ad absurdum? This problem arises in various ways in medical epistemology, ethics and philosophy of science. Solving it once and for all would be a great service to philosophy of medicine.

With regard to identifying the most interesting and important substantive problems I think that they are often found either where new technological developments open up new questions (or re-actualise old ones), or at the interface between philosophy of medicine and other areas of philosophy. It is impossible to provide an exhaustive or even representative list of these issues but let me mention four examples.

Telemedicine and the use of expert systems will force us to revisit our analysis of the clinical encounter and of what it means to make a clinical decision and be responsible for it.

The rise of "personalised medicine" will call upon us to think again about the old problem of how statistical data, for instance relative risk ratios at the population level can be properly interpreted and applied at the individual level.

Developments in the neurosciences and in philosophy of mind / neurophilosophy create new challenges for the philosophy of psychiatry.

And there are interesting overlaps between philosophy of medicine and political philosophy in the area of resource allocation in health care.

How do we decide which of these are most important? As mentioned above I personally "do" philosophy of medicine because I think it has important implications for health care so my ranking of importance would include three elements: 1) is the issue philosophically interesting, 2) is there any hope of saying something significant about the issue and 3) is it important for health care practice.

Here it is important to note that "important for health care practice" does not mean "perceived as important by health care professionals". That an issue is perceived as important by health care professionals will make it easier to communicate and collaborate with health care professionals, but there are many important issues that are not at least initially perceived to be important.

References

Holm S. The peacable pluralistic society and the question of persons. Journal of Medicine and Philosophy 1989; 13:379-86

Holm S. Does chaos theory have major implications for philosophy of medicine?. Medical Humanities 2002; 28: 78-81

Holm S. The phenomenological ethics of K.E. Løgstrup – a resource for health care ethics and philosophy?. Nursing Philosophy 2001; 2(1): 26-33

Woods J. Privatizing death: Metaphysical discouragements of ethical thinking. Midwest Studies in Philosophy 2000; 24: 199-218

9

Ingvar Johansson

Professor emeritus in theoretical philosophy

Department of Historical, Philosophical and Religious Studies; Umeå University, Sweden

---

1. Why were you initially drawn to philosophy of medicine?

I am a philosopher who has mainly been doing philosophical ontology and overarching philosophy of science, but I have never felt any inclination always to stay within these areas. In the early eighties, I was asked by a philosophically interested GP, Niels Lynøe, to join a discussion group where some GPs met and discussed common problems, often of a very theoretical character. Some years later he asked me to become assistant supervisor for his dissertation in social medicine (1991), and I accepted; also, he persuaded me that we should together write (in Swedish) an introductory book to the philosophy of medicine. So we did; it appeared in 1992, and got a second enlarged edition in 1997. Furthermore, I was supervisor for a Licentiate thesis (1995) that he wrote at the department for the philosophy of science, Umeå University.

Without Niels, now professor in medical ethics at Karolinska Institutet, Stockholm, I am sure I would never have tried to do anything that is specific to the philosophy of medicine. One reason for my assuredness has to do with the 'overlap' view that I say more about in my answer to question 2. In brief: medicine and philosophy overlap and do sometimes interfere with each other; philosophy of medicine cannot be reduced to the mere application within medicine of already outside of medicine established philosophical results. This view implies, among other things, that in order to do something within the philosophy of medicine you need either to have quite a knowledge yourself of medicine or to cooperate with someone who has. My collaboration with Niels has recently resulted in our book "Medicine & Philosophy. A Twenty-

First Century Introduction" (Ontos Verlag: Frankfurt 2008). Most of my answers below can be abstracted from this book.

2. What does your work reveal about Philosophy of medicine that other academics, citizens, or economists typically fail to appreciate?

It shows that there is an important overlap between medicine and philosophy. On part of medicine it involves all the main fields such as clinical practice, clinical research, biomedical research, epidemiological research, medical informatics, and medical ethics; on part of philosophy it involves not only epistemology and ethics, but also philosophical ontology. As the term 'overlap' makes clear, I am not of the opinion that medicine is through and through impregnated by philosophy. Often, philosophy is of no immediate consequence for specific research projects or for specific clinical situations, and then medicine and philosophy can fruitfully proceed in isolation from each other. But now and then a philosophical issue becomes highly relevant, and this possibility is of such a character that all medical scientists and practitioners had better to be aware of it.

Many practical-minded people and scientists think falsely that they cannot enter philosophical territory without making a jump over a fence that marks a border between philosophical reflections and their ordinary activities. But philosophical problems can pop up just where at the moment they are situated. Suddenly their implicit stance in relation to an unnoticed philosophical problem makes a difference to their normal undertakings (examples in my answer to question 5). In such situations, they tend to react by defending a certain specific philosophical position while at the same time denying that it is a philosophical position. All opposing philosophical views appear to them to be simply nonsense, and then there is of course no reason for them to engage in a dialogue.

The overlap between medicine and philosophy can pop up in class-room situations, too. Sometimes students put forward questions that the teachers evade by saying that they are 'too philosophical'. Of course, such an answer may in a particular situation be exactly the adequate one; there is a division of labor between science and philosophy. I am fairly sure, however, that many contemporary medical teachers use the phrase 'too philosophical' in such a derogatory sense that the students are given the false impression that philosophical reflections can never ever be of scientific or practical relevance. One consequence of the overlap view is that this is bad teaching.

Another important consequence of the proclaimed overlap is that neither medicine nor philosophy can be the utmost arbiter for the other. On the one hand, philosophers should not be given a juridical function within medicine, i.e., they should not be appointed legislators, judges, or policemen with respect to scientific theories, methodologies, and ethical problems within medicine; their role should only be that of a consultant. On the other hand, neither should medical people tell intervening philosophers to shut up only because they are philosophers. Quite another thing is that it might be relevant for medical people to ask philosophers better to learn what present-day medicine in fact says; the philosopher's abstract eye may easily miss some important details.

3. What, if any, practical and/or social-political obligations follow from studying medicine from a philosophical point of view?

I think a simple and brief analogy can make my basic position clear. In society at large, as I see it, citizens have two complementary obligations. First, they have to tell their community about serious crimes that they are fairly sure have been committed; but, second, they should not try to give a final verdict on the accused persons until these have been proven guilty. In medical communities, first, medical-philosophical persons have the obligation to make their community aware of what they take to be serious philosophical mistakes that influence medical research and/or clinical practice. But, second, they have to await discussion and scrutiny before they really try to make their views change medical research and/or clinical practice.

4. What do you see as the most interesting criticism against your own position in philosophy of medicine?

Let me first state my position, which can be dubbed 'pragmatic realism'. It has three main parts. First, the one already exhibited: there is an overlap between medicine (and science in general) and philosophy. Second, medical science and medical practice are, just like all human scientific and practical endeavours, fallible. Third, with respect to both moral and scientific-methodological rules I am a so-called particularist, i.e., I am convinced that substantial general rules of both these kinds can be overruled by a new situation, but that this fact does not imply moral and epistemological relativism. In other words: both moral and methodological particular decisions can be judged as being more or less right, but all substantial general moral and methodological rules are only default rules.

Of the three parts distinguished, the overlap view has been very little discussed for reasons that I will soon explain; fallibilism has been discussed, but not really in the form it takes when combined with particularism. In relation to morals, particularism goes back to Aristotle and his concept of 'phronesis' or 'practical wisdom'; and until recently it has very much been neglected in contemporary philosophy. I guess that it is this relative lack of criticism of pragmatic realism that explains why, at the moment, I do not really know what to regard as the most interesting criticism; today, I am equally convinced of all the three parts. But let me expand on what I have just said.

Both the overlap view and fallibilism have as their background and presupposition a belief in the traditional so-called correspondence conception of truth: if an assertion (a truthbearer) about something in the world is true or truthlike, then there is something in the world (a truthmaker) that corresponds or partly corresponds to the assertion and makes it true or truthlike, respectively. It is such correspondence truth-claims that are said to be fallible, and it is in their search for such truths that science and philosophy overlap. This truth conception, however, has been heavily under fire during the second part of the twentieth century; a fact which partly explains the lack of direct criticisms of the overlap view and fallibilism. Most prominent among the critics of correspondence is the Oxford philosopher Michael Dummett, who thinks it does not make sense to speak of a distinction between truthbearers (assertions or propositions), truthmakers (facts), and an external relation (correspondence) between them; instead facts are claimed to be identical with true propositions. He takes his departure in the philosophy of mathematics, and his anti-realist position may even in my opinion be adequate in relation to mathematics, but it is impossible to generalize and say that it is adequate also in relation to the empirical sciences. A denial of the correspondence conception of truth is one thing in relation to mathematics and quite another in relation to medical science.

Such anti-realism aside, the overlap view is surrounded by three opposing views. There are ontological realists who place philosophy above science, others place it below, and some beside. In the 'above' camp we find Kant and all pure rationalists such as Descartes and Hegel. They claim not only that all philosophical problems can be solved independently of the sciences, but also that empirical science has to stay within a framework discovered by philosophy alone. In the 'below' camp we find those who think that true philosophy should confine itself to logic and conceptual analysis; a position most conspicuously stated by the logical positivists, but adhered to also by some other strands within analytic philosophy. These thinkers can be said to place philosophy below science, since they think that philosophy can only contribute to knowledge about the world by sharpening the logic-conceptual tools used in empirical science. Then, third, there are a few philosophers in the 'beside' camp. They claim that philosophy is of no relevance whatsoever to science; most famous is the epistemological anarchist P. Feyerabend. However, if there are few philosophers in this camp there are the more scientists. Many scientists seem to be happy to agree to what the Nobel laureate physicist Richard Feynman is reported to have said: 'Philosophy of science is about as useful to scientists as ornithology is to birds'.

All the three overlap-opposing views make philosophers completely sovereign in philosophy; correspondingly, the 'below' and the 'beside' view make scientists completely sovereign in science, whereas the 'above' view of Descartes-Kant-Hegel subordinates theoretical scientists to philosophers. At present, however, none of these views is on a broad scale discussed within the philosophy of science. Rather, the overlap view is made invisible by the epistemologically nihilistic view of radical social constructivism, i.e., the view that both everyday conceptions and all scientific theories are, just like novels and plays, only social constructions without truth content. But I cannot take this view seriously. The claim 'everything is a social construction' does, when thought through, (i) deny modern cosmology and evolutionary biology, and (ii) break the semantic rule that it does not make sense to speak of a construction if there is no constructor outside of the construction. The radical social constructivists do simply not take these oft repeated remarks seriously. Neither do they seriously consider what fallibilism has to say about 'theories as social constructions', which means that they do not clearly see that there is another competing position beside theirs that regard theories as a kind of social constructions, too.

Fallibilism is the view that we can never, not even in empirical science, be absolutely certain that we have obtained truths about the world. As far as I can see, this view has today become the so to speak natural epistemological position among natural and medical scientists. It differs from skepticism in being affirmative, claiming that it is incredible to think that we have no knowledge at all; especially in view of all the science-based technological and medical inventions that have revolutionized the world. And it differs from radical social constructivism in claiming that certain kinds of social constructions, especially empirical-scientific theories, can have truth-content. Fallibilism was first explicitly spelled out by the pragmatist Charles Sanders Peirce (whose truth-conception differs from most other pragmatists such as W. James, J. Dewey, and R. Rorty) and the critical rationalist Karl Popper (whose views are not identical with everything that goes under the label 'critical rationalism'). However, both Peirce and Popper, each in their own way, combine fallibilism with other views in such a way that it can be hard to see the essence of fallibilism. Some interpretations and criticisms of it are not really concerned with fallibilism as such. Let me briefly explain.

Despite his fallibilism, Peirce puts forward a kind of criterion of truth: true is what in the long run the scientific community will unanimously regard as being true. This is future social consensus around correspondence, but many modern pragmatists see only the consensus aspect, forget about correspondence, and turn Peirce's fallibilism into a social constructivism. Popper claims that there are no truth criteria within science, and has not been misunderstood the way Peirce has. But, in contradistinction to Peirce, he combines his fallibilism with a belief in the existence of general methodological rules and a criterion for what makes a theory scientific, i.e., he combines fallibilism with non-particularism. Now, it seems to me as if many social constructivists take the correct criticism of Popper's methodology and falisifiability criterion to be also a criticism of his fallibilism, but this is a mistake.

Particularism in moral philosophy has recently, mainly in the hands of J. Dancy, become a position that is seriously discussed in analytic moral philosophy, but not in the philosophy of science. I am quite convinced, however, that what particularism claims about the non-existence of ethical principles applies to methodological principles, too. Aristotle, the first particularist, constrained his particularism to morals and politics, but it has to be extended even to science. Such a proposal has been put forward before, but mainly within the hermeneutic philosophical tradition. And since hermeneutic philosophers have, to put it mildly, great qualms in accepting the correspondence conception of truth, they do not propose exactly the combination of scientific-theoretic fallibilism and scientific-methodological particularism that I believe in.

I would like to end this answer with some speculations about the views of contemporary medical scientists. I have got the impression that most of them implicitly endorse both fallibilism and particularism, but deny the overlap view. This has a peculiar effect: they think of themselves as having no philosophical position at all, and that the combination of fallibilism and particularism is not a philosophical position. Therefore, they regard themselves as being pragmatists in a completely non-philosophical sense of this term. But this means only that they behave as fallibilists and particularists without trying by argument to defend these positions. Their defense is: 'I am not a philosopher; I am just a pragmatic person'. Nonetheless, implicitly and inevitably, they have philosophical positions. Of course, my hope is that all medical scientists will realize the overlap between medicine and philosophy, and then become explicit defenders of a philosophical pragmatic realism.

5. With respect to present and future inquiry, how can the most important problems concerning Philosophy of Medicine by identified and explored?

As can be seen from my last answer, I think there are no general rules by means of which one can identify such problems; neither how to proceed in order to solve them. But I will be happy to point at two medical-philosophical problems whose solutions I think would mean much to the development of medicine and, by the way, to philosophy, too. One is how to understand so-called psychosomatic phenomena. Can they once and for all be deemed social illusions comparable to the natural-perceptual illusion that the sun is moving over the sky? Or, if not: how should psyche-to-soma causation be conceived? Although causal talk is ubiquitous in both everyday life and scientific life, the notion of causality is philosophically elusive. The other problem I would like to highlight is how to interpret singular probability statements of the following two forms: (a) 'this particular person runs, with the probability p, a risk of getting disease D', and (b) 'given the treatment T, there is a probability p that this particular patient will be cured'. Now some more words about each problem.

According to what philosophers use to call 'folk psychology', there are many phenomena that can be given the abstract philosophical label 'psychosomatic phenomena'. For instance, to say 'his strong will saved his life' is to imply that a psychic will-to-live was a causal factor in the curing of a deadly somatic disease; and to say 'his new promotion seems to have made his medical problems disappear' is to imply that certain somatic problems disappeared because of a happy psychological mood. And this folk psychology is as alive in medical research as it is in everyday life. Here, however, the psychic cause in the psyche-to-soma causation is not said to be a positive psychological mood or a will to become cured, but the patients' psychic expectations that that they will be cured. As normally conceived, the dummy pills of randomized controlled trials are assumed to function because the persons in the control group expect to become cured by them; the placebo effect is assumed to be a psyche-to-soma effect. The very aim of the RCTs is of course to isolate and find a purely biomedical effect, a soma-to-soma causation, but this does not alter the fact that the meaningfulness of the RCTs themselves presupposes the existence of psyche-to-soma causation. That is, psychosomatic phenomena are at one and the same time both accepted and disregarded. I find this is odd.

However, for several reasons it is not easy to study psychosomatic phenomena; and some of these reasons are philosophical. Philosophy has so far not managed to reach a consensus about either how to define the essence of psychic phenomena or how to understand causation; in particular, not psyche-to-soma causation (which in contemporary analytic philosophy is discussed under the label 'mental causation'). Surely, the search for causal relations can always start with a search for correlations, but as long as it is unclear what constitutes a psychic phenomenon, even presumed psyche-soma correlations can be questioned. This does not mean that I am of the opinion that medical researchers who want to study psychosomatic phenomena have to await philosophical developments made by pure philosophers; it means only that such researchers should realize that their hypotheses are not philosophically innocent.

It would, I guess, involve quite a change in medical research if all hitherto assumed placebo effects should become regarded as being either mere statistical illusions or due to spontaneous purely biomedical curing. And, let it be noted, the remarkable helicobacter pylori success story in relation to peptic ulcer cannot be regarded as having finally settled the question of the existence of psychosomatic phenomena. In fact, the effect of antibiotic treatments of peptic ulcer was studied by means of RCTs with their psychosomatic notion of placebo effects. It is one thing to show that one specific assumed kind of psychosomatic phenomenon was an illusion, quite another to show that there can be no kinds of such phenomena at all.

Let me now turn to the other medical-philosophical problem that I find important: the interpretation of medical singular probability statements. This problem is, as far as I can see, even more neglected than that of psychosomatic phenomena.

Both laymen and psychiatrists speak of certain persons as having a 'suicidal tendency'. Sometimes the judgment is given the form of a vague probability statement, for instance: 'there is a probability of about 1/6 that Joe will commit suicide'. Such a probability statement is not about how probable it is that the speaker knows that Joe will commit suicide, i.e., it is not an 'epistemic-subjective' probability statement. But neither is it at first sight about a relative frequency in the world (a 'frequency-objective' probability statement). At least formally, the statement is only about Joe and a property he has, a tendency to commit suicide and the probability of its realization; it can be called a 'singular-objective' probability statement. To my mind, many medical people do not care to hold these three kinds of probability statements distinct, and I think there are some practical reasons behind this lack, but these can in the theoretical context at hand be disregarded. In what follows I will put epistemic-subjective probability statements aside, and focus only on the distinction between frequency-objective and singular-objective probability statements.

In contemporary philosophy it is very common to regard tendencies as ontologically impossible entities, and I have got the impression that something similar is true in medical research. On such a presupposition, first appearances notwithstanding, the statement 'there is a probability of about 1/6 that Joe will commit suicide' is not about a tendency Joe has, but about a relative frequency in a population or set to which Joe can be ascribed membership. No doubt, many singular probability statements have to be interpreted as being no more than short-hands for statements about a relative frequency. For instance, the formally singular-objective statement 'the probability that this lot will be a winning lot is 1/6' is only short-hand for the frequency-objective statement 'the relative frequency of winning lots in this lottery is 1/6'. Let me use a thought experiment in order to show what such an interpretation of tendency statements would imply.

Assume a certain community where the relative frequency of

suicides in the community as a whole is 1/6000, but that for people more than fifty years old it is 1/4000 and for males 1/5000. This means that for a male person over fifty (call him Joe), all the three following statements are true when they are interpreted as short-hands for relative frequencies of suicides among (a) citizens in general, (b) citizens older than 50, and (c) male citizens:

(a) 'the probability that Joe will commit suicide is 1/6000'
(b) 'the probability that Joe will commit suicide is 1/4000'
(c) 'the probability that Joe will commit suicide is 1/5000'

But Joe cannot have all these three probabilities as real singular-objective tendencies; this would be like weighing 70, 80, and 75 kg simultaneously. At a certain moment, a suicidal tendency can have only one specific strength and probability to be realized. If, as positivism and much other contemporary ontological thinking claim, all non-epistemic singular probability statements whatsoever have to be short-hands for relative frequencies, then, really, psychiatrists ought to stop talking about suicidal tendencies. But there is another philosophical possibility, one that I think should be seriously explored: to take the psychiatrists' spontaneous notion of 'tendency' at its face value and claim that, literally, there are tendencies in the world. Furthermore, if humans can truly be ascribed tendencies, then many medical singular 'risk-of-getting-disease' judgments may be interpreted as being about tendencies, too.

Think of the statement 'the risk that Joe will get the disease D is 1/3'. Even if this formally singular probability statement is as a short-hand for a relative frequency statement true, as a substantial tendency statement it may be completely false; and false in two different ways. On the one hand, Joe may have no tendency or propensity at all to get the disease D; he is simply for some reason immune to D. On the other hand, he may have a tendency towards the disease that is so strong that sooner or later he will inevitably get it. But, third, the statement may be true both as a short-hand frequency-objective statement and as a singular-objective tendency statement. If so, then Joe is bearer of a real propensity/tendency and an accompanying risk to get disease D, but for some reasons there is only a probability of 1/3 that the propensity/tendency will become realized.

Hopefully, this last answer of mine can shed light also on my answer to question two: there is an important overlap between medicine and philosophy.

10

Niels Lynøe

GP and professor of medical ethics; Director of Stockholm Centre for Healthcare Ethics, Karolinska Institutet, Stockholm, Sweden

---

1. Why were you initially drawn to Philosophy of Medicine?

It was not until my second year of medical studies I became interested in the philosophy of medicine. It was in 1974 at Copenhagen University. If I had not attended this new optional course in philosophy of medicine, I would probably have discontinued my medical studies and returned to my original idea of becoming a computer scientist. You may say that the philosophy of medicine saved my future commitment to medicine. Before the course I had not the slightest idea of what medicine or medical studies were all about. And for the first two years I was no less confused.

As indicated, I was not primarily interested in medicine. The reason why I actually attended medical school was that the medical faculty arranged a preparatory course of physics, chemistry and mathematics. It was arranged so as to enable arts students to attend medical school. But in my case, passing the course also allowed me to study computer science, which was my main reason for attending the course. During the course, however, I met some interesting and nice people. After the course they encouraged me to go on studying medicine and why not? If it was interesting I could continue and if not, I could just quit. Accordingly I chose to accompany my friends.

For the first three years we did not see a living patient, but we saw quite a lot of dead ones in connection with anatomy and the dissection course. At that time the content of the first two years of medical studies included basic disciplines like chemistry, physics, statistics, genetics, histology, psychology etc. The closest these disciplines came to clinics was illustrated in an exercise in physics: we were asked to calculate the weight of a newborn baby carried by a midwife in an elevator when the (rather old-fashioned) balance showed x kilos and the elevator was accelerating at a certain speed (I think the exercise was supposed to illustrate Einstein's equivalence principle). Not even the course in medical psychology had any connection with clinical skills; it was about fishes' (carps') behaviour under certain conditions, perhaps supposed to illustrate the reasoning of behaviourism. Most students understood these courses as mere initiation rites and not as something preparing us for our future work as physicians. Or was it perhaps a method of probing our motivation for entering the medical profession? Many medical students were actually so strongly motivated that they were prepared to learn the telephone directory by heart if ordered to. Unfortunately (or luckily), I didn't have that kind of motivation.

I remember some discussions with teachers – discussions which might tacitly have prepared my interest in philosophy of science. A chemistry teacher, for example, maintained that he was mainly dealing with acids and bases and since these entities could not be left- or right-winged, he as a chemist could neither be left- nor right-winged. Accordingly, he thought he had actually proved that he as a scientist was completely value-neutral. I also remember a (male) teacher of embryonic anatomy who, while making drawings on the blackboard, small-talked about the development of human species. Purely in passing he mentioned that males were more developed than females: 'males climbed down from the trees before females' he said. When some female students protested loudly, accusing him of propagating gender-prejudices, he was somewhat astonished and exclaimed: 'What I am saying is nothing but scientific facts'. In this manner we leant something about scientists' spontaneous philosophy. Before having read Thomas Kuhn I also got the impression that scientific observation might be both theory- and value-loaded.

As a second-year student in medical school I was not only confused, I became more and more frustrated. Accordingly it was a great help that the medical faculty had decided to offer an optional course in philosophy of medicine. After having passed the course I got an idea of what medicine and particularly what the study of medicine was about. During the course we discussed what a sick person is, disease labelling issues, what is understood by causality in medicine, what is understood by scientific knowledge, the role of tacit knowledge, scientific and ethical reasoning etc. I discovered that philosophy was interesting, but I did not find it interesting in itself – to me philosophy combined with medicine was interesting. And vice versa: medicine was interesting, but not in itself – medicine combined with philosophy was interesting. And ever since that time I have been more or less addicted to this combination.

Later on, when we attended the clinical part of the study, we met quite a lot of physicians who were supposed to function as role-models in patients' encounters. I will never forget a clinical demonstration of a patient suffering from RA in front of approximately 200 medical students. The teacher did not speak to the patient, he was so occupied with his teaching that he did not even notice that the patient felt uncomfortable. The teacher was rather curt when he bent the patient's in arm and leg joints, and obviously he was hurting her. Most students felt uncomfortable too. But nobody said anything – not even after the patient had left the theatre. Eventually I initiated several studies about the ethics of clinical training and medical teachers as role-models. Today I wonder whether these studies are compensations for not reacting adequately when the RA patient during my clinical training was so harmed and wronged.

We also attended an open surgery department, to which all patients from the central part of Copenhagen arrived. The young surgeons were skilled and taught us how to stop bleedings and how to sew open wounds etc. After the course the chief surgeon asked us to evaluate it. Even though we had learnt a lot we had also noticed that several surgeons were rather unkind and violent towards the patients. The chief surgeon reacted with a rhetorical question: 'Wouldn't you prefer to be treated by a skilled but unkind surgeon rather than being treated by a nice but unskilled one?' At that moment we were not able to provide an adequate answer. Not until the end of our medical study did we react and write a letter to the editor of the Danish Medical Journal, questioning whether bad encounters were a logical or an ontological necessity (1). Even though the myth about being either skilled and unpleasant or unskilled and pleasant is rather old-fashioned, it might still be alive – compare the TV soap about Dr House. The myth has followed me all through the years and actually influenced the initiation of several empirical projects about patient encounters.

During my medical studies several events and experiences prepared and trigged my interest in the philosophy of medicine. But these events and experiences have also influenced and are currently influencing the direction of some of my research. I am currently happy that I chose to follow the same road as my friends. "Non, je ne regrette rien".

2. What does your work reveal about the Philosophy of Medicine that other academics, citizens, or economists typically fail to appreciate?

Philosophers of medicine who have a philosophical background usually acknowledge a critical and analytical approach when conducting research within the area. Traditionally, philosophers are even slightly sceptical to whether empirical research might contribute to the philosophy of medicine. Accordingly, few philosophers have competence in empirical research. A large part of the research in which I have been involved has had an empirical approach. It has been possible to develop projects and make observations as well as interpret empirical data in a way that few philosophers might have done on their own. Epidemiologists might, however, also have difficulties in understanding and discussing the philosophical aspect of empirical results. I will mention two examples which might illustrate in what manner my research might have contributed to increased knowledge in philosophy of medicine. But before providing such examples, I would like to say a few words about and the reasons behind my empirical approach.

In 1979, after completing my medical studies, I moved to Sweden and began my clinical training in order to become a GP. The end of this training found me in a department of public health with a tradition of quantitative empirical research. I was fairly free to do whatever I found interesting, and I began co-operating with a philosopher, Ingvar Johansson – acknowledged elsewhere in this book. In my department I was made responsible for teaching the philosophy of medicine and research ethics. Eventually I wrote my thesis in public health with Ingvar Johansson as my supervisor. This was the 'beginning of a long and beautiful friendship' and co-working. One of our first projects was to write a textbook in philosophy of medicine, now in its third, English edition (2). I think that this co-work and the academic freedom within the department of public health shaped my approach to the philosophy of medicine.

During the years I have initiated several empirical studies regarding informed consent and patients' participation in clinical research as well as in the clinical training of medical students. I have also initiated studies about physicians' and the general public's reasoning, e.g. about loyalty conflicts and end-of-life decisions. In order to clarify the reasoning and the argument of highest priority we have used clinical cases (vignettes) in which significant values and principles are at stake.

One of these studies concerns physicians' and the general public's attitudes and reasoning about physician-assisted suicide (PAS). Apart from the presentation of the results of the attitudes – 35 % of the physicians were pro and 39% against, whereas in the general public 75% were pro and 10% against – we also asked the participants to evaluate and prioritise between different pros and cons (3). The main arguments pro were respect for the patients' autonomy and minimisation of suffering at the end of life. The main arguments against PAS were that the non-maleficence principle should overrule the autonomy principle and legalising PAS might jeopardise trust in health-care. An additional argument against PAS was that patients at the end of life do not know their own good – a classical paternalistic argument. Very few of the physicians supported this latter argument, indicating that physicians are not paternalistic or that physicians know that overt profession of paternalistic attitudes is not politically correct. In order to examine whether physicians actually entertain paternalistic attitudes, albeit in disguise, we scrutinised the empirical data more closely. We observed a paradoxical phenomenon: Psychiatrists were the group most in favour of PAS, but they did not support the most common pro-argument – the autonomy argument. The construction of the questionnaire and the priority setting of arguments allowed us to create an index of disguised paternalism. By means of the paternalism index we found that psychiatrists and oncologists showed a significantly more disguised paternalist attitude than GPs and surgeons. In order to validate the results we also constructed an autonomy index and found that GPs and surgeons scored significantly higher in the autonomy index compared to psychiatrist and oncologists (4). Since it is difficult to defend a paternalistic attitude openly, we think that such indirect and case-dependent studies say more about disguised paternalism than studies where the issue is studied directly.

From the same study of physicians' reasoning about PAS we also identified another aspect which a pure epidemiologist or a pure philosopher probably would not have been able to. We asked the physicians what might happen to their own trust in the health-care system if PAS were to be legalised, and finally we asked what they believed would happen to confidence among the general population. The response options were in both cases that trust would 1) decrease, 2) not be influenced or 3) increase. We found that the group of physicians opposed to PAS could not distinguish between their estimations of their own reactions and their estimations of the reactions of the general populations. Those of the physicians who were pro or undecided were able to make this distinction. Those against PAS also used the most colourful language when commenting the questionnaire. From a rational point of departure we might expect the general population to appreciate their autonomy being empowered in the end of life, e.g. by legalising PAS. But it was only physicians pro PAS and those in doubt who made their estimations in accordance with rational thinking and actually close to how the general population actually answered. The study indicated that strong value-based attitudes influence estimations of future events (5). It might be discussed whether observations and estimations are both theory- and value-impregnated. Particularly estimations of consequences of future events seem to have great impact on ethical reasoning.

Philosophers might enjoy it when they succeed in making a proper analysis or developing a convincing argument. I also enjoy these aspects of philosophical reasoning. But looking at empirical data for the first time fills me with an almost childish happiness. Accordingly, I can't tell how much I appreciate the combination of new empirical data and good philosophical arguments.

3. What, if any, practical and/or social-political obligations follow from studying medicine from a philosophical view?

Generally, philosophers of medicine are used as experts in different settings such as research ethics committees, ethics boards at hospitals and universities and also national ethics boards. In this manner they might have a great impact on decision-making in specific cases as well as on policy-making. But even as researchers in the philosophy of medicine they might influence the societal discussion of particularly ethical issues. Although some of these ethical issues might be difficult to discuss publicly, the general public and media are actually interested in discussions of, e.g., the end-of-life issues. The above mentioned studies about physician-assisted suicide (PAS) yielded some quite surprising data. Almost 35% of Swedish physicians were positive and there was no majority against PAS. The publication of these results brought about a public discussion about PAS. But the public discussion became particularly heated when the National Board of Medical Ethics one year later (2008) launched a proposal of a democratisation of different end-of-life issues. The Swedish Minister of Social Affairs also joined in this discussion and almost banned his own board's suggestion and request for a public investigation. As a member of the Christian Democratic Party, he was strongly against PAS, and even though the proposal also contained suggestions about palliative sedation, nothing happened in the political setting.

Recently a discussion about the values within palliative care, with special reference to sedation-therapy, has taken place in the Swedish Medical Journal. Even though the ethical reasoning seems not to influence or change the palliative care physicians' attitudes, palliative care values are actually at variance with Swedish healthcare legislation. As an example of such discussions I think there is a story worth telling from the Swedish setting, in which legal and philosophical analysis run together and point in the same directions.

According to the European Association of Palliative Care (EAPC), palliative care physicians should be restrictive about offering sedation therapy to patients at the end of life (6). The goal of palliative care is neither to prolong nor to shorten life. Accordingly, sedation therapy should not be offered until the last twenty-four (or perhaps forty-eight) hours of life. Sedation therapy in itself is not supposed to shorten life. But if sedation therapy is combined with not providing the patient with fluids, this combination might shorten life if sedation therapy is initiated, say, a week before expected death. During the last twenty-four or forty-eight hours sedation therapy in combination with no fluid would not hasten death. This is the reason why palliative physicians are restrictive with sedation therapy, at least when treating competent patients.

This point of view, however, is controversial – and not only ethically speaking. According to Swedish healthcare laws, physicians ought to provide adequate treatment to the patient at the end of life despite the fact that such treatment might shorten life. If the patient's suffering is perceived as unbearable and refractory to conventional symptom-relief, physicians have an obligation to offer sedation therapy. And even if such a patient is expected to live for some weeks or months longer, the physician should offer the patient sedation therapy. If a physician abstains from offering such a patient sedation therapy with reference to an arbitrary time-limits, it is actually considered as neglect (7).

According to the EAPC task force documents, it is the patients who evaluate the symptoms as unbearable or tolerable, but it is the physician who determines whether or not the symptoms are refractory. This point of departure is not consistent: if unbearable suffering is something the patient determines, why should the physician be in a better position to determine whether or not symptoms are refractory (8)? The only point seems to be that it is the physician who finally determines whether or not sedation-therapy should be offered.

In a Swedish setting, palliative physicians offering sedation-therapy usually distinguish between the dying patient being competent or incompetent. An incompetent patient suffering from, e.g., a brain-tumour is usually sedated several weeks before expected death, if the patient's symptoms are refractory. The competent patient, however, is not offered sedation therapy until the last 24 or perhaps 48 hours. According to a Swedish palliative physician, the reason for treating competent and incompetent patients differently is that autonomy counts (9). According to the physician the patient's autonomy should be protected and preserved as long as possible. Palliative care physicians reason that sedation therapy will destroy autonomy and accordingly should not be offered. If the patient's autonomy is already lost, there is nothing to protect and accordingly sedation therapy is acceptable. If a competent patient wants to get sedated in order to discontinue unbearable suffering, the physician might refuse it with reference to protecting the patient's autonomy. But protecting a competent patient's autonomy and not respecting it could amount to nothing but a sophisticated kind of disguised paternalism: paternalism in the name of autonomy (10).

Another kind of disguised paternalism is maintained by the same palliative physicians when they claim that there are medical contraindications for sedation therapy. They maintain that during sedation there is risk that patients might, for example, develop infections, thrombosis and muscle atrophy (11). It is rather difficult to understand what the point is, Thirty years ago physicians were also restrictive about providing morphine to dying people because of the risk of developing substance dependency. What is this, if not paternalism in disguise?

Today it is more or less well-known that medical technologies should be evidence-based. But it is also known that treatments should be value-based. Evidence-based treatment that is not value-based is of no value if, for example, the patient is not accepting it. In medicine it is important to examine values held by patients in order to see whether or not they will accept a new treatment. Palliative care seems to be rather value-based and focused not primarily on the patient's values but rather on those of the palliative care physicians. Since nobody else seems to focus on these issues, highlighting such values and controversies is an important task for philosophers of medicine.

To me it is obvious that patients' autonomy should be respected even at the end of life. If physicians think that patients' autonomy should be protected rather than respected at the end of life, they should provide convincing arguments for that view. If such arguments are lacking we should respect, rather than protect autonomy.

It is well-known that the suicide rate is significantly higher among patients suffering from progressive neurodegenerative diseases such as Huntington's disease and ALS (12-14); this is also the case regarding prostatic cancer and probably other forms of cancer too. I would like to offer the following hypothesis: If sedation therapy is offered on a more patient-centred and less physician-centred basis, suicide rates among the above mentioned group of patients will decrease.

4. What do you see as the most interesting criticism of your position in the philosophy of medicine?

For a long time now I have noticed that physicians who are interested in philosophy are not considered as real physicians – even if you are still seeing patients. Such a physician is considered as too theoretical and is usually not taken seriously in debates. Lately I have also recognised that philosophers who are interested in biomedical issues are referred to as 'blood and sperm' philosophers – and this is not meant as a complement. A pure philosopher is supposed to deal with meta-ethics/philosophy - not blood, sperm and other real-life entities. I am not sure whether a 'blood and sperm' philosopher is taken less seriously when debating, compared to 'pure' philosophers.

From my own perspective I am considered neither as a pure physician nor as a pure philosopher. Accordingly, nobody is supposed to listen to what I am saying, even though or particularly if, I have performed a good philosophical analysis or provided reasonable arguments. This is actually a paradox: if as a philosopher you make a good analysis and provide proper arguments, few within the medical field will listen to you. If you are a chief surgeon with great external recognition, many will listen to you, even if your reasoning is based on gut feelings and on poor arguments or none at all. And if as a philosopher you dare to say something about the rules of proper reasoning, the reaction will not fail to appear – indicating that you are not supposed or in a position to determine the rules of debate (15-16).

This is an interesting issue, since physicians at least are supposed to develop a professional attitude towards other specialities within clinical medicine. A professional attitude implies that you are aware of the limitations of your knowledge and competence. Accordingly, we might expect physicians to be wary when trespassing in other areas such as the philosophy of science and ethics. But in such cases we actually observe the opposite reactions: consultant physicians think that they are automatically experts on medical ethics and medical scientists think that they are automatically experts in the philosophy of science and research ethics. And they think that they are experts due to their long experience of clinical practice and medical research.

According to my experience this is not quite true, but neither is it quite wrong. It is not quite true because many senior clinicians are actually not able to explain or answer, e.g., medical students' or patients' questions about ethics. Many clinicians maintain that the students have to do this or that because it accords with the routines of the clinical department. In many cases no discussion or reasoning is offered, and if reasoning actually is provided even the students recognise it as poor. Relatively few medical researchers are able to provide a meta-perspective on their own research activity. If a researcher actually talks about his or her standpoint on the philosophy of science, it often disagrees with what they actually do as a researcher.

To a certain extent the experienced clinician or the experienced researcher might in a sense be referred to as an expert in clinical ethics or the philosophy of science and research ethics. An experienced clinician has seen many patients and participated in many difficult situations involving ethical conflicts. In many cases they have resolved such conflicts by using their intuition or just following old routines. If all parties concerned are satisfied we might say that an ethical problem has been solved without ethicists, which might indicate that the experienced physician actually does have ethical expertise or that this expertise is not necessary for solving ethical problems. Sometimes physicians would maintain that there was no ethical aspect – it was a purely clinical decision.

But in cases where the experienced physician does not succeed in solving the ethical conflict it becomes obvious that we need more specific ethical competence. Ethical intuition is not enough and other competences are demanded: ability to analyse and indentify relevant actors, ethical principles and values at stake. In other words, what is demanded is competence in the grammar of ethics.

The experienced physician might have learnt medical ethics by adopting traditions and values more or less automatically by having observed older (good and bad) role-models. In other words the experienced physician has learnt ethics the natural way, just like a person learning a new language on his own. It is difficult to maintain that the experienced physician is not able to solve ethical conflicts, but he/she is not able to identify the ethical argument and which values were in conflicts with each other. He/she is neither able to specify what he/she actually did in ethical terms if or when he/she succeeded in solving a problem. The experienced physician's competence is merely based on tacit knowledge, and in this respect he/she might be characterised as an expert. The other kind of expert in medical ethics is based on knowledge of ethics grammar. Similar aspects might be applied when discussing research and philosophy of science (2).

These aspects are not to be considered as criticism of a specific position in the philosophy of medicine, but they are rather critical for the philosophy of medicine in general, at least when working in a medical faculty. Although it is rather tiresome, I suggest a strategy of patiently repeating over and over again why the philosophy of science is needed. And it is needed since we are living in a changeable world, where a fallibilistic view is necessary both within the clinical settings as well as in the research settings. It is rather uncommon for problems, conflicts or dilemmas to be soluble by means of old routines and clinical intuitions. When patients and even medical students are asking why clinicians are doing this or that, they are no longer satisfied with answers like: it is routine and we have always done it in this way.

As can be seen from my experiences of the philosophy of medicine, it is not a specific position that is being defended, I am defending the existence and relevance of the philosophy of medicine in a medical setting. I am not sure whether I think that this fundamental criticism is interesting in a philosophical sense.

5. How can the most important problems concerning philosophy of medicine be identified and explored?

Together with a colleague I recently wrote a paper in the Swedish Medical Journal entitled 'Ten myths about medical ethics to be put to an end', in which we commented on different prejudices about medical ethics (17). A physician had his own explanation why nobody listened to representatives from medical ethics (18). He thought medical ethicists had no authority since there were no sanctions if someone transgressed ethical rules or guidelines. The critical physician expected that ethicists should function as police, prosecutor and judge in one person. He thought that nobody would pay attention to ethical guidelines, rules or reasoning until ethicists were inclined to take such a position.

The question is whether it is possible or even desirable for a medical ethicist or philosopher to act as police or prosecutor. I think it is possible but not desirable. Ethical reasoning is supposed to follow some logical rules and ethicists are supposed to react when researchers and physicians contradict themselves. There are several rhetorical tricks, guilt by association, defective slippery-slope arguments, abuse of reductio ad absurdum arguments, use of euphemisms and dysphemisms and not being consistent or acting against supervenience criteria. For an ethicist it would be fairly easy to criticise ethical reasoning and act like an ethics police. But the desirability of so doing is open to question. What might happen if we acted in such a direction? Would more clinicians and researchers develop respect for logical and ethical rules? Would they improve their moral reasoning and assume greater moral responsibility towards the patients and their colleagues? And the question is whether medical ethicist and philosophers would like to become a logical or moral police – how convenient would that be?

One problem for the philosophy of medicine is that certain kinds of issues and research within the area is not solely supportive of clinical practice or medical research. The goal of the philosophy of medicine is to critically, analytically and empirically scrutinise the healthcare system and medical research. Accordingly, the results or analyses might sometimes become provocative and inconvenient to both clinicians and medical researchers. Studying the quality of informed consent in clinical research is not popular if informed consent procedures are not optimal. Studying unofficial values among physicians is not popular if a group of physicians support special values. And according to some physicians we should avoid controversial issues like end-of-life decisions and patients' participation in decision-making. Even the abortion issue is now and then critically discussed in Sweden, though in connection with prenatal diagnostic procedures. Sometimes I personally feel that some physicians would like me to hold my tongue. But I think that ethicists and medical philosophers have an obligation to participate particularly in uncomfortable discussions. It is not necessary to participate as police or prosecutors, but critically analyse arguments and facilitate reasoning.

Another area where the 'police' function might be relevant is in medical schools. For example, medical students have a special attitude towards the ethics or philosophy of science, at least when a course in the discipline is not examined. Students themselves are aware that they are taking courses with no examination less seriously than courses which are actually examined. Is it possible to exist as a discipline within the medical curriculum without performing an examination, if all other courses actually are examined? In order not to disavow the discipline and not to make it a pedagogical anomaly, I think that it is important that the philosophy of medicine be examined in the same way as every other medical discipline within the medical curriculum.

A special problem with the philosophy of medicine is, however, that many physicians and researchers do not know what the discipline is actually about. Most of the senior physicians and researchers have received no instruction in clinical ethics, research ethics, philosophy of science etc. They might react rather differently: some might think it is very important and have positive expectations, others might think that it is waste of time and have negative expectations. Depending on their attitude or prejudices, they will act or not act in a certain direction.

In many medical schools the curriculum has been changed, at least in Europe. Some of the classical preclinical disciplines have had their space reduced in order to accommodate new disciplines and new elements in the medical curriculum. If the philosophy of medicine is introduced under such circumstances, it might also bring about problems. Even though teachers from other areas do not openly denigrate the philosophy of medicine, they tell students that their (more important) disciplines have been squeezed out in order to make room for the philosophy of medicine. In other words, other teachers' negative attitudes might inform the students' attitudes towards the philosophy of medicine (19). This is a big problem, and again we have to patiently explain what the philosophy of medicine is able to contribute. Even if the medical student is not aiming at medical research we might as a minimal demand expect that the future physician become a reflective practitioner (20). And in order to become a reflective practitioner you need to know something about ethical and scientific grammar.

There are certainly plenty of tasks for future research within the philosophy of medicine, regarding new medical technologies, e.g. nanotechnology and analogy or similarity-thinking in relation to nature. It is of the greatest importance to study the new technologies' impact on human beings, current values, our understanding of normality, health, pathology, causality, disease-labelling etc. It is also important to focus on research regarding the tacit dimension of knowledge.

One special area of interest for the philosophy of medicine is experimental philosophy. It is a new way of conducting empirical research of more specific philosophical issues such as causality and intentionality. I think that Knobe's experiments about intentionality are of great importance for ethical reasoning when discussing, e.g., the principle of double effect (21). Contrary to what we might expect, consequences, not intentions, are relevant when the general population and even physicians assess the morality of an act. (22)

References

1) Lynøe N, Hegeler I. Is it a logical or ontological necessity to be an unkind physician? *Ugeskrift for Læger.* 1979;

2) Johansson I, Lynøe N. *Medicine & Philosophy – A Twenty-First Century Introduction.* Ontos Verlag, Frankfurt 2008.

3) Lindblad A, Löfmark R, Lynøe N. Physician assisted suicide: A survey of attitudes among Swedish physicians. *Scand J Public Health.* 2008;36:720-7.

4) Lynøe N, Juth N, Helgesson G. How to reveal disguised paternalism? *Medicine, Health Care and Philosophy.* 2010;13:59-65

5) Lynøe N, Juth N. Do strong values influence estimations of future events? *J Med Ethics.* 2010;36:255-6.

6) Cherny NI, Radbruch L. European Association for Palliative Care (EAPC) recommended framework for the use of sedation in palliative care. *Palliative Medicine* 2009;23(7):581-93.

7) Leijonhufvud M, Lynøe N. Sedation-therapy that shorten life – homicide or adequate medical treatment? *Lakartidningen* 2010;107 (45): 2772-3.

8) Juth N, Lindblad A, Lynøe N, Sjöstrand M, Helgesson G. European Association for Palliative Care (EAPC) framework for palliative sedation: an ethical discussion. *BMC Palliat Care.* 2010 Sep 13;9:20.

9) Eckerdal G. *Sedation in palliative care – the doctor's perspective.* In Tännsjö T. (Edt) *Terminal sedation: euthanasia in disguise?* Kluwer Academic Publishers. Dordrecht, 2004.

10) Sjöstrand M, Eriksson S, Juth N, Helgesson G. Paternalism in the name of autonomy". J Medicine and Philosophy. Forthcoming.

11) Engström I, Eckerdal G. Time limits are needed for palliative sedation. *Lakartidningen.* 2010;107: 3297.

12) Baliko L, Csala B, Czopf J. Suicide in Hungarian Huntington's disease patients. *Neuroepidemiology.* 2004;23:258-60.

13) Fang F, Valdimarsdóttir U, Fürst CJ, Hultman C, Fall K, Sparén P, Ye W. Suicide among patients with amyotrophic lateral sclerosis. Brain. 2008;131:2729-33.

14) Fang F, Keating NL, Mucci LA, Adami HO, Stampfer MJ, Valdimarsdóttir U, Fall K. Immediate risk of suicide and cardiovascular death after a prostate cancer diagnosis: cohort study in the United States. *J Natl Cancer Inst.* 2010 3;102(5):307-14.

15) Lynøe N. Also the ethical debate must follow certain rules! *Läkartidningen* 2009;106:832-3.

16) Milberg A, Karlsson M. Lynøe shall not determine how we should discuss medical ethics.
http://www.lakartidningen.se/07engine.php?articleId=12076

17) Engström I, Lynøe N. Ten myths about medical ethics to be put to an end. *Lakartidningen* 2010; 107 (40):2419-21.

18) Bodegård G. The powerlessness of medical ethics. *Lakartidningen.* 2010;107:2658.

19) Lynöe N, Juth N, Helgesson G. Case study of a framing effect in course evaluations. Medical Teacher. Forthcoming.

20) Schön D. *The reflective practitioner – how professionals think in action.* Basic Books; New York, 1983.

21) Knobe J, Nichols S. *Experimental Philosophy.* New York: 2008, Oxford University Press.

22) Juth N, Tillberg T, Lynöe N. Intentions in critical clinical settings: study of medical students' perceptions. J Med Ethics. 2011;37(8):483-6.

11

Ruth Macklin

Professor of bioethics
Albert Einstein College of Medicine, New York, USA

---

1. Why were you initially drawn to Philosophy of Medicine?

My answer to this question requires a bit of biographical history. When I entered graduate school in the mid-1960s, philosophy of medicine did not exist. Philosophy of science was the king of disciplines. Debates raged in philosophy journals about the nature of scientific explanation, whether explanation in biology fit the model of explanation in physics, whether explanation in the social sciences was essentially different from the natural sciences. In the graduate program in which I was enrolled, there were many courses in "Philosophy of....": philosophy of physics, philosophy of mathematics, philosophy of language, philosophy of biology, philosophy of social science, philosophy of history, philosophy of art, and philosophy of law. Interestingly, however, there was no course in philosophy of medicine. I was interested in philosophy of psychology, and wrote my doctoral dissertation on what was then a hot topic: theory of action. This was not really philosophy of psychology, however. The inquiry was purely one of analytic philosophy, divorced from what psychologists were studying and writing about. Theory of action addressed such questions as: "What is the difference between "raising my arm" and "my arm going up"? What is the role of intentions in explanation of action? How do intentions to act differ from motives to act? Following the paradigm of explanation in physics, can human actions be explained by the so-called "hypothetico-deductive model"? My doctoral thesis made no reference to any empirical research or even to theories put forth by psychologists. Our philosophy of science heroes were philosophers from the Vienna Circle and the Berlin Group, and contemporary philosophers such as Karl Popper, Thomas Kuhn, Carl Hempel, and Ernest Nagel.

After receiving my PhD degree, when I joined the philosophy faculty, I decided to turn to some genuine psychology to see what empirical research looked like and how psychologists actually explained human behaviour. This led me to the writings of B.F. Skinner and other behaviourists, as well as to the writings of Sigmund Freud, Carl Jung, and others in psychiatry whose educational background was in medicine rather than social science. They also led me to the book by Thomas Szasz, The Myth of Mental Illness. These writings were a world apart from the abstract notions of the philosophers whose works I studied when researching theory of action. At the same time, my heart (as well as my mind) remained in philosophy, and I became interested in concepts of health and disease, especially mental health and mental illness. This interest formed a bridge between analytic philosophy and the field of medicine, and I found myself doing philosophy of medicine—probably without fully realizing it and without calling it by that name.

By the early 1970s, the fledgling field of bioethics (then called 'medical ethics') was emerging. Along with colleagues from the philosophy department where I was teaching at the time, I worked on developing curricular materials for a course called Moral Problems in Medicine, we co-edited the first anthology in the field (with the same title), and I began to publish articles on various topics in the field then still called 'medical ethics'. What drew me to the field was the intersection of philosophy and real-world concerns. Interested since my youth in social and political issues, but also having my intellectual home in academic philosophy, I found a perfect fit in moral problems in medicine and health policy. In 1976 I left the Philosophy Department at a university and was employed at The Hastings Center, a bioethics research organization, and soon thereafter at a medical school. For the past thirty years I have remained on the faculty of the same medical school.

2. What does your work reveal about Philosophy of Medicine that other academics, citizens, or economists typically fail to appreciate?

Many academics in philosophy have considered bioethics to be an inferior discipline. "Applied" philosophy has, for some, had a status lower than "pure" philosophy, somewhat in the way physical scientists tend to look down on the field of engineering. For many years, most philosophy departments at the most prestigious universities in the United States included no faculty members who worked solely or mainly in philosophy of medicine or bioethics. Although that has since changed, it remains true that many prominent philosophers hold themselves apart from—or above– those of us working in applied fields. An example occurred several years ago when a legal case in medicine came before the United States Supreme Court. Many of us were contacted by attorneys to sign on to amicus curiae briefs supporting one side or the other in the case. There was a "bioethicists' brief," which included many philosophers, and a separate "philosophers' brief." The latter group, apparently, did not want to be listed as part of the former group. I believe that my own work and that of many other philosophers in the field have retained the philosophical rigor our education required of us.

Interestingly, academics from other fields in the humanities and the social sciences have had the opposite reaction to work in bioethics. Some historians, anthropologists, and sociologists have criticized philosophical writings in bioethics as being too abstract, "acontextual," that is to say, analytic and lacking empirical grounding. However, since bioethics is a multidisciplinary field, physicians, social scientists, and legal scholars are here to contribute the contextual material that may be lacking in some of the literature written by philosophers. Other critics, at least where bioethics is concerned, come from medicine itself. There are still physicians (though not as many as in the early decades) who largely reject philosophers working in the field and teaching in medical schools for "not having been in the trenches." This military analogy suggests that one can speak or write knowledgeably about something only if one has had direct practical experience. That viewpoint, at worst, is anti-intellectual, but also contains a kernel of truth. Philosophers who work entirely within an academic setting in a philosophy department and who never enter the real world of medicine often lack relevant knowledge or experience that renders them hopelessly naive.

With regard to economists, there are indeed some who care about the value of things as well as their cost. However, the dominance of cost-benefit analysis as the economists' mode of reasoning typically leaves little room for the other important human values that exist in the world of medicine. As for citizens, it is impossible to generalize. One of the books I wrote was published by a trade house rather than an academic press. Writing in typical philosophical (academic) fashion, I sprinkled the text with quotations, which I then commented on. The editor from the trade publisher who was assigned to my book repeatedly told me: "We don't care what those guys think; we care about what you think." Although that book did receive some favourable reviews, one reviewer from a major U.S. city newspaper wrote: "A dry exercise, best left on the shelf." That judgment surprised me, as every chapter was filled with medical cases drawn from my experience as a teacher and consultant in a large, urban medical centre. However, this episode did reveal the obvious fact that many ordinary readers (and book reviewers) do not appreciate a philosophical approach.

3. What, if any, practical and/or social-political obligations follow from studying medicine from a philosophical point of view?

This is an interesting question. I'm certain there are those who would answer "none." For such people, academic study is primarily an intellectual exercise, from which nothing follows for practical, social, or political life. My own view is that several obligations follow from studying medicine from a philosophical point of view; but I, like many others, may fail to fulfil these obligations in every instance. The chief obligation is to make clear to the public in general what problems exist and how they can be understood and analyzed. It is most unfortunate that the main channels for getting medical information to the public are through media sound bites. Over the years, I have frequently been interviewed by newspaper reporters, as well as radio and TV journalists. In virtually all cases, my remarks are either edited down to a sound bite, and in many other cases are taken out of context. A rare counterexample was two very long interviews by the U.S. public television journalist, Bill Moyers. A highly intelligent, well-informed interviewer, Moyers allowed me to speak in paragraphs, on a range of topics, and the two television shows that emerged from the interviews lasted one hour apiece. The only other avenue for clarifying complex issues in medicine for the public is speaking engagements open to general audiences. At one such event, years ago, I was explaining the role of Infant Care Committees in hospitals after a law was passed mandating such committees. A visibly pregnant woman sitting in the front row protested: "You mean some hospital committee is going to decide what will happen to my baby?" I was able to explain that such committees act on the presumption that the parents are the ones who should make such decisions, unless their decisions are "clearly against the best interest of the infant." However, I'm not sure the woman was fully convinced that her authority to decide on behalf of her baby would not be snatched from her by the bureaucracy.

Dangers may lurk, however. It is not only a philosophical analysis of a problem in medicine that we may bring to the public; it may also be a position that takes an ethical stance on a controversial issue. Especially on a radio or television news program, journalists seek opposing points of view in order to bring some theatre to listeners or viewers. This is normally not objectionable, since as we often say in bioethics, "reasonable people may disagree." However, a phenomenon exists on U.S. "talk radio" in the form of nasty, right-wing political extremists who host the programs. They are not interested in soliciting reasoned points of view. Rather, they seek only to attack a moderate or liberal interviewee who does not adhere to the extreme position taken by the host. As a strongly pro-choice person, I was vilified on the one occasion I was foolish enough to accept an interview on one of those talk-radio shows. Nothing I said could clarify, explain, or justify a pro-choice position to listeners, given the rude behaviour of the radio host.

Sometimes an attempt to reach the public may fail, through no lack of effort. Several years ago a colleague and I were seeking to mount a project to study the inner workings of the U.S. Food and Drug Administration (FDA). Information had come to light about the agency's failure to pull a drug off the market when postmarketing evidence had already revealed the product's unexpected and dangerous side effects. Other information from newspapers indicated that serious conflicts of interest existed within FDA and among physicians who served as consultants or members of the agency's scientific committees. My colleague, a health economist, and I wrote an article for the editorial opinion pages that was rejected by two leading U.S. national newspapers. We wrote a proposal to fund our proposed project that was rejected by several philanthropic foundations. In the end, we just gave up, and the failings of the FDA have been reported in occasional newspaper articles, but without the depth and rigor that we had hoped to use in our project.

Beyond the obligation to bring the results of philosophical study of medicine to the public, there may also be obligations of an activist sort. Although such opportunities are not always available to academics, a variety of possibilities are suggestive. One possibility is to join with activist groups working to seek changes or reforms of existing medical practices that one has analyzed or studied. A simpler option is to contribute money to an organization that provides medical services or is an advocate for public policy. In my case, such an organization is Planned Parenthood of America, to which I regularly contribute. My systematic study of reproductive health in the U.S. and abroad has revealed many unmet needs, and the anti-woman agenda of the administration of George W. Bush has contributed to difficulties that women in the U.S. have encountered in obtaining reproductive health services.

A somewhat different channel is to agree to dedicate time and effort to governmental, nongovernmental, or international committees or agencies that require the expertise of volunteers or consultants, who are typically paid a very low fee. Among the examples of this time and effort are the following. The U.S. Federal government and state governments have had a variety of temporary or permanent committees devoted to medicine and health. Philosophers have been invited to serve as members of such committees or alternatively, to help staff the committee. Well-known U.S. committees have been the National Commission for the Protection of Human Subjects of Biomedical and Behavioural Research, the President's Commission for the Study of Ethical Problems in Medicine and Biomedical and Behavioural Research, and the President's Advisory Committee on Human Radiation Experiments. Another possible role is that of service as a reviewer for an agency or organization that makes research grants to individuals or institutions. I have served as a reviewer for the U.S. National Institutes of Health, which is a time-consuming activity that pays very little for my time, and have reviewed also for the U.K. Wellcome Trust. I currently serve as an adviser in bioethics on two committees at the World Health Organization, which pays no honoraria for service. Similarly, I have served on two committees of the Joint United Nations Programme on HIV/AIDS, which also does not pay for service. I have been an unpaid member of committees at the Institute of Medicine of the U.S. National Academies of Science. I consider it an obligation to serve on such committees, an obligation that flows from the expertise I have gained as a scholar in philosophy of medicine for more than thirty years.

4. What do you see as the most interesting criticism against your own position in philosophy of medicine?

I have stood for many positions—not just one—during my long years as a philosopher working in bioethics and philosophy of medicine. In my book, Against Relativism: Cultural Diversity and the Search for Ethical Universals in Medicine, I argued that ethical universals exist and provide a basis for ethical criticism of traditional practices that either harm people, treat them unjustly, or both. Examples I discussed are female genital mutilation, denying women safe, legal abortions, and preventing adolescent girls from having access to contraceptives, among others. Ethical relativists are many and vocal, among philosophers, students, and the general public. I am not certain that ethical relativists' criticisms are the most interesting, but they have probably been the most vocal against positions I have taken. Those criticisms have never been sufficiently persuasive to lead me to weaken my defense of ethical universals.

In recent years, I have done a good deal of research and writing about multinational research, typically sponsored by industrialized countries or the pharmaceutical industry, and carried out in developing countries. I have taken strong positions in published articles and in my book, Double Standards in Medical Research in Developing Countries, arguing that sponsors, along with other stakeholders, have an obligation to make successful products of research available to research participants, if they still need them when their participation ends, and also to the resource-poor community or country where the research is conducted. There are legions of critics of my position among philosophers, other bioethicists, and researchers. I am not certain that the criticism of this position I have taken in multinational research is the most interesting, but it has been the most difficult one for me to respond to.

Another position I have taken in the area of medical research is opposition to the use of placebo controls in most studies of new products when a proven medication already exists. A few philosophers have been among those who criticize my opposition to placebo controls, but most criticism of that position comes from industrial sponsors of research, officials at the Food and Drug agency that regulates pharmaceutical products, and researchers themselves. Sponsors prefer placebo-controlled medical trials because they are quicker and cheaper than studies comparing a new product with an existing one. In addition, it is far easier to demonstrate that an experimental drug is superior to a dummy pill than to show that it is superior to another company's product. The drug regulatory agency prefers placebo-controlled trials because officials believe the results are clearer and easier to interpret than head-to-head trials of two drugs. And researchers choose placebo-controlled trial designs for the same reasons as sponsors, but also because the FDA requires them in most cases. My position is that it is unethical to deny to half the people in a research study a beneficial product they could receive if they were not in the trial. This often comes down to a debate between bioethicists and research methodologists. In research design, the gold standard is a placebo-controlled trial, where the participants are randomly chosen either for the active intervention arm of the trial or the placebo group. Bioethicists have a different gold standard, one that requires doing the least harm possible in research. Since denying a group of participants a treatment proven to be beneficial is likely to cause more harm than providing them with the treatment, the placebo-controlled trial is less ethically acceptable than one that compares the experimental drug to an existing drug. Critics do have a point, however, when they say that the experimental drug may turn out to be harmful, so in that case the placebo group is better off. Nevertheless, since the safety of the experimental product has always been tested in smaller studies before this design with placebo is conducted, there is much less chance of finding the experimental medication to be harmful at this stage. What makes the debate interesting is that it pits a methodological gold standard against an ethical gold standard in medical research.

A variation of this debate has occurred in the context of the type of international research discussed above. Some people have taken the position that it is ethically acceptable to use placebo controls in a developing country where no beneficial product is available outside the research study, even when such products do exist in the sponsoring country. In cases where a placebo-controlled trial could not ethically be conducted in the industrialized country, such as a study of a new AIDS medication, I have taken a strong position against placebo controls in the developing country. I find this type of "double standard" ethically unacceptable. Criticism of my position in this case may be the most interesting, in one sense, because of circumstances in which it occurred. Sharp disagreement erupted among members of a small writing group of which I was a member, appointed to draft a revision of international ethical guidelines for biomedical research. It is one thing for philosophers to take a position in an article or a book they publish. Reviewers of the article or book, or other readers, can mount a criticism in another published article. Similarly, when a philosopher makes a presentation at a meeting or conference, a commentator or member of the audience can criticize the position and the speaker can reply. However, when members of a group charged with writing guidelines or preparing a consensus document disagree, there must be a resolution or else the guideline or document does not get written. In the case of the writing group of which I was a member, we reached an impasse. The only solution was to take the matter out of the hands of the writing group and arrive at a procedure whereby the impasse could be overcome. The result was a compromise, of sorts, which left me reasonably satisfied but my opponents in the argument very unhappy.

Some criticisms of positions I have taken are ones that I have actually come to adopt. As I reflect on my own changes in positions I once held, I recognize that I came to abandon my former views not as a direct result of criticisms, but rather upon further thinking and the evolution of my views. Two examples involve monetary payments to people. The first is payment to research subjects for their participation. In an article I wrote in the early 1980s, I expressed concerns about payments being "undue inducements" to participate in research. I still believe that in truly risky research, participants should not be paid very large sums of money. But I find myself arguing with colleagues on the research ethics committee in my institution about payments to subjects in social and behavioral research. Such research is typically very low risk, so there is little or no danger that people will be induced into doing something they are really opposed to but agree to do only because of the money. The second case is that of payment to women who donate their eggs to other women who are infertile. Although I did not argue that such payments were strictly unethical, I did say that the practice was "unsavory" as it amounted to commodification of the human body. I now believe that women who donate eggs deserve some compensation for their time, discomfort, and the inconvenience involved in being an egg donor. A related topic is payment to women to serve as a surrogate for another couple's baby. In the 1980s I served on a committee in the state of New Jersey, established after a prolonged episode in which a woman who served as a surrogate refused to turn over the baby to the couple that had commissioned her services. The committee debated furiously about the ethics of paying surrogates, and at the time I sided with those who argued against payment. I now believe that payment to reproductive surrogates is not unethical for basically the same reason that payment to egg donors is not unethical. What is interesting about the criticism of my previously held positions is that I came to believe in the very criticism of the view I formerly held.

5. With respect to present and future inquiry, how can the most important problems concerning Philosophy of Medicine be identified and explored?

The answer to this question depends on what one assesses to be the most important problems. Even within bioethics, a sub-area of philosophy of medicine, the great diversity of problems opens numerous different avenues. Many of these are already identified, whereas other are probably not yet identified. A very brief list of very big problems already identified will illustrate. Access to adequate and affordable health care is a huge problem for 46 million people in the United States who lack health insurance, as well as for the majority of the population in resource-poor countries. Although scholars in bioethics have explored and continue to study this problem, at least in the U.S. the problem remains lack of political will by policy makers, and not lack of scholarly exploration of the problem. In developing countries the problem is less that of political will than of very limited financial and human resources. Many studies have addressed the so-called "10-90 gap" in funding for health research: only ten per cent of worldwide expenditure on health research and development is devoted to the problems that primarily affect the poorest 90 per cent of the world's population. This is a severe imbalance that is evident and has been studied, but solutions are not readily forthcoming. Another problem that has been extensively explored is the health disparity within and among countries between the financially well-off and the poor. Middle- and upper-income people are healthier and live longer than their poorer counterparts even within industrialized countries. Other big problems already identified are those on which there is unresolved and probably unresolvable controversy: pro-choice and anti-abortion public policies; and aid-in-dying, physician-assisted suicide, euthanasia. One wonders whether there is anything further to be explored by philosophers or others regarding these problems.

Some problems in bioethics have arisen only when something new emerges, thereby giving rise to a new set of issues to explore. HIV/AIDS, a new disease, emerged in the early 1980s, and was accompanied immediately by areas requiring urgent study: how to mitigate discrimination against those who became infected; what precautions were needed to prevent healthcare workers from becoming infected; how best to keep the blood supply safe; how to conduct ethically acceptable research into the disease; how to allocate healthcare resources to infected individuals; what obligations exist across national boundaries, to name only a few. Another relatively new area in medicine is the explosion of knowledge and associated technology in human genetics. As new techniques to detect genetic disorders have been developed, and as more and more genetic disease markers are discovered, the problems grow apace. How many genetic diseases should newborn screening encompass? What is the obligation to disclose genetic information to people when the meaning of that information is as yet unknown? How to protect confidentiality of genetic information that may be harmful if revealed to employers or insurers? What obligation, if any, do physicians have to disclose significant genetic information that may affect family members of their patients? And a fundamental question: should genetic information be deserving of special protections, resulting in "genetic exceptionalism" in medicine? Then there was the scientific breakthrough in the ability to derive pluripotent stem cells from human embryos. That development has led to a whole new area of exploration: the ethics and politics of stem cell research; once again, debates over the moral status of human embryos; what sorts of restrictions are needed for stem cell research to proceed ethically; what is acceptable by way of creating human-animal chimeras for stem cell research? An even newer scientific development, nanotechnology, has already given rise to the field of "nanoethics." Like "genethics" before it, nanoethics has come under fire as an artificially created field of study. One article refers to its proponents as "trying to midwife a new discipline." As all of these examples illustrate, the problem does not lie so much in identifying or even exploring areas of study, but rather in setting priorities among them.

Yet problems do exist for philosophers seeking to explore problems in philosophy of medicine, including bioethics, when they need grant money to help support their salary. Even in relatively rich countries, there is generally much less grant money to support the work of philosophers than that of scientists. In the U.S., for example, the National Institutes of Health has a large budget for research, but money for philosophers only if they are conducting empirical studies, which most philosophers are not equipped to do. The Fogarty International Centre of the NIH supports training programs in research ethics, so philosophers can obtain financial support if they direct a training program; but training programs are not research programs to identify and explore ethical problems in research. The one U.S. government program that has supported projects in philosophy of medicine is the human genome program within the NIH. The Ethical, Legal and Social Implications (ELSI) Research Program was established in 1990 as an integral part of the Human Genome Project (HGP) to foster basic and applied research on the ethical, legal and social implications of genetic and genomic research. Philosophers have conducted conceptual, theoretical, and analytic studies, as well as projects in bioethics and collaborative work with others on empirical studies.

Some areas in bioethics have been very much neglected till recently, and even now, continue to attract a minority of philosophers. I am referring here to topics in global medical research and health. Interest in international problems in medicine and health has lagged far behind other mainstream topics, such as end-of-life issues, organ transplantation, and other topics in clinical medicine. A look at the conference programs of the American Society for Bioethics and Humanities over the years reveals a distinct lack of panels and presentations devoted to issues beyond the borders of the United States. This is rather striking, given the creation and expansion of Global Health Centres in schools of medicine and public health in the U.S.

As I said in the beginning of my answer to this question, there is great diversity in beliefs about what are the most important problems in the area of bioethics. From the beginning of this field of inquiry, some scholars have focused on clinical aspects of modern medicine. A smaller number tackled problems in health policy, but that area has grown considerably over the years. When AIDS emerged, some bioethicists began to focus largely on HIV-related problems, others have always done most of their research on topics in human reproduction, and recent years have seen the growth of public health ethics. Scholars rarely have difficulty identifying new areas of inquiry, as the rush toward "nanoethics" indicates. The biggest difficulty, as I see it, is moving from scholarly exploration of big problems to implementing or even trying to arrive at solutions. We philosophers still sit in ivory towers, even if our offices are located in academic medical centres. It is left to policymakers—and unfortunately, politicians—to implement even well-thought out, informed, practical solutions to problems in philosophy of medicine traditions.

References

Aghina, M.J. (1978): Patiëntenrecht. Een kwestie van gewicht. Van Gorcum, Assen.

Anspach, R.R. (1993): Deciding who lives. Fateful choices in the intensive-care nursery. University of California Press, Berkeley.

Arnold, R.M. and Forrow, L. (1993): Empirical research in medical ethics: an introduction. Theoretical Medicine 14: 195-196.

Beauchamp, T.L. and Childress, J.F. (1983): Principles of biomedical ethics (2nd ed.). Oxford University Press, New York/Oxford.

Berg, J.H. van den (1969): Medische macht en medische ethiek. Callenbach, Nijkerk.

Bleuler, Eugen (1921): Naturgeschichte der Seele und ihres Bewustwerdens. Eine Elementarpsychologie. Springer, Berlin.

Bosk, C.L. (1979): Forgive and remember. Managing medical failure. The University of Chicago Press, Chicago/London.

Bosk, C.L. (1992): All God's mistakes. Genetic counseling in a pediatric hospital. The University of Chicago Press, Chicago/London.

Bosk, C.L. (2008): What would you do? Juggling bioethics and ethnography. University of Chicago Press, Chicago.

Brody, H. and Miller, F.G. (1998): The internal morality of medicine: Explication and application to managed care. Journal of Medicine and Philosophy 23(4): 384-410.

Caplan, A.L. (1983): Can applied ethics be effective in health care and should it strive to be? Ethics 93; 311-319.

Daniel, S.L. (1986): The patient as text: A model of clinical hermeneutics. Theoretical Medicine 7: 195-210.

Dijk, P. van (1978): Naar een gezonde gezondheidszorg. Ankh-Hermes, Deventer.

Engelhardt, H.T. (1974): Explanatory models in medicine: Facts, theories, and values. Texas Reports on Biology and Medicine 32 (1): 225-239.

Gordijn, B. and ten Have, H. (eds.) (2000): Medizinethik und Kultur. Grenzen medizinischen Handelns in Deutschland und den Niederlanden. Fromman-Holzboog, Stuttgart − Bad Cannstatt.

Gordijn, B. (2003): Die medizinische Utopie. Eine Kritik aus ethischer Sicht. Nijmegen.

Have, H. ten (1980): Wijsbegeerte der geneeskunde. Algemeen Nederlands Tijdschrift voor Wijsbegeerte 72: 242-263.

Have, H. ten (1980): Self-help. Backgrounds, postulates and problems of a new phenomenon. Huisarts en Wetenschap 23: 305-308.

Have, H. ten (1983): Geneeskunde en filosofie. De invloed van Jeremy Bentham op het medisch denken en handelen. De Tijdstroom, Lochem.

Have, H. ten (1984): Ziekte als wijsgerig probleem. Wijsgerig Perspectief 25: 5-12.

Have, H. ten & van der Arend, A. (1985): Philosophy of medicine in the Netherlands. Theoretical Medicine 6: 1-42.

Have, H. ten (1990): Verleden en toekomst van medische filosofie. Scripta Medico-philosophica, Schrift 7, p. 5-18.

Ten Have, H. (1994): The hyperreality of clinical ethics: A unitary theory and hermeneutics. Theoretical Medicine 15: 113-131.

Have, H.A.M.J. ten (1995): Medical technology assessment and ethics. Ambivalent relations. Hastings Center Report, 25:13-19.

Have, H.A.M.J. ten (1998): Images of man in philosophy of medicine. In: Evans, M. (ed.): Critical reflection on medical ethics. JAI Press, Stamford (Conn.), p.173-193.

Have, H.A.M.J. ten and Lelie, A. (1998): Medical ethics research between theory and practice. Theoretical Medicine and Bioethics 19: 263-276.

Have H.A.M.J. ten and Welie J.V.M. (eds.) (1998): Ownership of the human body. Philosophical considerations on the use of the human body and its parts in healthcare. Dordrecht, Boston, London: Kluwer Academic Publishers.

Have, H. ten (2000): The zapping animal. Oscillating images of the human person in modern medicine. In: A-T.Tymieniecka and Z. Zalewski (eds.): Life - The human being between life and death. Analecta Husserliana LXIV, Kluwer Academic Publishers, Dordrecht, pp. 115-123.

Have, H.A.M.J. ten (2001): Theoretical models and approaches to ethics. In: H.A.M.J.ten Have & B. Gordijn (eds.): Bioethics in a European perspective. Kluwer Academic Publishers, Dordrecht/Boston/London, pp. 51-82.

Have, H. ten (2004): Ethical perspectives on health technology assessment. International Journal of Technology Assessment in Health Care 20(1): 71-76.

Have, H.A.M.J. ten (2005): A communitarian approach to clinical bioethics. In: C. Viafora (ed): Clinical bioethics. A search for the foundations. Springer, New York, pp. 41-51.

Have, H. ten and Welie, J. (2005): Death and medical power. An ethical analysis of Dutch euthanasia practice. Open University Press, Maidenhead (UK).

Illich, I. (1975): Medical nemesis. Calder & Boyars, London.

Jonsen, A.R. and Toulmin, S. (1988): The abuse of casuistry. University of California Press, Berkeley, CA.

Kuczewski, M.G. (1997): Fragmentation and consensus. Communitarian and casuist bioethics. Georgetown University Press, Washington, D.C.

Leder, D. (1988): The hermeneutic role of the consultation-liaison psychiatrist. Journal of Medicine and Philosophy 13: 367-378.

Lelie, A. (1999): Ethiek en nefrologie. Dissertation, University of Nijmegen.

McKeown, T. (1976): The role of medicine. Dream, mirage or nemesis? The Nuffield Provincial Hospitals Trust, London.

Newton, A.Z. (1995): Narrative ethics. Harvard University Press, Cambridge (Mass.).

Payer, L. (1988): Medicine & Culture. Varieties of treatment in the United States, England, West Germany, and France. Henry Holt and Company, New York.

Pellegrino, E.D. (1976): Philosophy of Medicine: problematic and potential. Journal of Medicine and Philosophy 1 (1): 5-31.

Pellegrino, E.D. and Thomasma, D.C. (1993): The virtues in medical practice. Oxford University Press, New York/Oxford.

Pellegrino, E.D. (1995): The limitations of empirical research in ethics. Journal of Clinical Ethics 6: 161-162.

Poirier, S. and Brauner, D.J. (1988): Ethics and the daily language of medical discourse. Hastings Center Report 18: 5-9.

Potter, Van Rensselaer (1971): Bioethics. Bridge to the future. Prentice-Hall, Englewood Cliffs, New Jersey.

Reinders, J.S. (1993): Over de medische praktijk als uitgangspunt van ethische reflectie. In: F.W.A. Brom, B.J. van den Bergh & A.K.Huibers (eds.): Beleid en ethiek. Van Gorcum, Assen, pp.226-232.

Sandel, M.J. (1996): Democracy's discontent. America in search of a public philosophy. The Belknap Press of Harvard University Press, Cambridge (Mass.) and London.

Sporken, P. (1969): Voorlopige diagnose. Inleiding tot een medische ethiek. Ambo, Bilthoven.

Stevens, M.L.T. (2000): Bioethics in America. Origins and cultural politics. The Johns Hopkins University Press, Baltimore and London.

Svenaeus, F. (1999): The hermeneutics of medicine and the phenomenology of health. Steps towards a philosophy of medical practice. Linköping University, Linköping (Sweden).

Thomasma, D. (1997): Bioethics and International Human Rights. Journal of Law, Medicine & Ethics 25: 295-306.

Toulmin, S. (1982): How medicine saved the life of ethics. Perspectives in Biology and Medicine 25: 736-750.

Tronto, J.C. (1993): Moral boundaries. A political argument for an ethic of care. Routledge, New York/ London.

Verbrugh, H.S. (1978): Paradigma's en begripsontwikkeling in de ziekteleer. De Toorts, Haarlem.

Willigenburg, T. van (1993): Ik ben een ethisch ingenieur! In: F.W.A. Brom, B.J. van den Bergh & A.K.Huibers (eds.): Beleid en ethiek. Van Gorcum, Assen, pp.189-204.

Wulff, H.R. (1980): Principes van klinisch denken en handelen. Bohn, Scheltema & Holkema, Utrecht.

Zaner, R.M. (1988): Ethics and the clinical encounter. Prentice-Hall, Englewood Cliffs, N.J.

Zola, I.K. (1973): De medische macht. Boom, Meppel.

Zussman, R. (1992): Intensive care. Medical ethics and the medical profession. The University of Chicago Press, Chicago/London.

12

Lennart Nordenfelt

Professor of Philosophy of Medicine
Dept of Medicine and Health Sciences, Linköping University, Sweden

---

1. Why were you initially drawn to Philosophy of Medicine?

My interest in philosophy of medicine started growing around 1970. At that time I was still heavily involved in completing my doctoral dissertation, which dealt with a topic in the philosophy of the humanities, more precisely the explanation of human actions (1974). My concern there was primarily with explanations in historical research. Thus, my scholarly interest still remained far from medicine.

However, I was at the time looking for a new line of research for my future life in philosophy. The concept of disease came to my mind as something that could be highly interesting and perhaps also original. At that time I knew of no previous work in this area. This certainly showed my great ignorance, but I can perhaps be excused. The modern philosophical discussion on medical matters had hardly begun in those days.

Why disease? A salient reason was that I was brought up in a family of doctors. My father was a doctor; so are two of my brothers and one sister-in-law. I have often started discussions and even quarrels with them about diseases and criteria of diseasehood. I was rarely content with their answers. My relatives are not particularly philosophically minded so our exchanges of opinion hardly ever became fruitful for me or ever went on for long. But my interest in the general topic was aroused and I thought the concept was worth exploring. I was also encouraged by the belief that the characterization of the concept of disease might be a completely new field for philosophical research.

My serious research in the area started around 1977, but before that I initiated a minor research project (pursued in my free time) where I interviewed some local family doctors about their criteria for sick-listing people. My own pre-understanding of the situation was strange. I thought that quite a number of people were continually simulating and just playing ill in order to get sickness compensation in an easy way. I was sincerely concerned and wondered how the family doctors could master this situation and stop the imposture. I still cannot remember why I had this prejudice about people. It is far from my present understanding of the situation.

I was intrigued when in 1975 I realized that the Zeitgeist had caught up with me. I noted the launch of the Journal of Medicine and Philosophy and very soon thereafter of Metamed (a German periodical which was eventually transformed into the well-known Theoretical Medicine and Bioethics). I wrote some early papers on the concept of disease where I was in fact arguing in the direction of a normative, holistic understanding of disease. Very soon, however, I came upon Christopher Boorse's early crucial articles (1975 and 1977) and I immediately turned a fan of his. I wrote a Swedish book on the semantics of the notion of disease where I followed Boorse's line of thought quite closely. Fortunately, this book was never translated into English.

There were three factors that heavily contributed to forming my mind about health and disease in the direction it took during the 80s and where it has remained ever since. First, my friend and colleague the Finnish philosopher Ingmar Pörn, with whom I had been discussing philosophy since 1970, also started to take an interest in the theory of health and disease. In our conversations he was the first to contemplate the idea of looking upon health as the ability to achieve goals. That was a completely new thought for me and it made me look upon the whole area from a new angle. Second, and perhaps surprisingly, the WHO initiative of proposing a classification of disabilities and handicaps, the ICIDH (1980), made me realize that my action-theoretic background could be of great use in the characterization both of the general concept of disability and of specific disabilities. From this insight it was not a long step to seeing how the concept of ability could play a major role in the characterization of health itself. Third, and this was perhaps the most decisive factor, I started reading Georges Canguilhem's outstanding work On the Normal and the Pathological (1943, English translation 1978) and I was greatly impressed by his analysis. The fact that he was a medical doctor and a distinguished theorist of biology contributed to my conviction that a normative analysis of the medical concepts must be the most reasonable one.

2. What does your work reveal about Philosophy of Medicine that other academics, citizens, or economists typically fail to appreciate?

As I indicated above, most, but indeed not all, of my writings in philosophy of medicine have been influenced by my action-theoretic outlook. The concepts of action and ability lie at the heart of my analysis of disability and handicap as well as my analysis of health. From health, which I define as a person's ability to realize his or her vital goals, given standard or otherwise reasonable circumstances, I derive the negative medical concepts, such as illness, disease, impairment and defect (1995; 2001). Ideas about explanations of action are central to my analysis of the activity of health promotion and of the different ways to influence health promotion (2000). Action explanation is also crucial for certain areas of psychiatry. Recently 2007 I have extended my action-theoretical analysis to the notion of workability in the content of health insurance.(argued that many mental illnesses partly involve behaviour or action under compulsion). Compulsion requires, in my opinion, a very specific theoretical analysis of action determination.

My general outlook is certainly not unique. Quite a few philosophers of health now have an action-oriented outlook on health. This is perhaps most salient in the case of K.W.M. Fulford (1989 and later). But I may differ from my colleagues in that I attempt to present action-theoretic concepts and analyses on a fairly detailed level and in a rather systematic way. The risk I run is perhaps that not everything in my analysis is relevant to the ultimate medical purposes.

Why is the action-theoretic platform essential? I think that it is crucial to underline—and this was perhaps more evident before the contemporary discussion about the nature of health had started—what should be the ultimate purpose of medicine and health care. The purpose is not in itself, I think, to cure the person's disease and bring his or her body (back) to a "normal" physiological state of affairs. The purpose is—and that is indeed normally the reason why people approach the health care system—to help people (back) to a thriving life where they are able to do what is important for them to do: work, socialize with their family and friends and cultivate their dearest interests. Thus, I believe, the notion of ability must have a crucial place also in the conceptual characterization of health. This certainly does not mean—and about this I have been quite clear in my writings—that I underestimate the biological knowledge about the human body or the psychological knowledge about the human mind. Once we have been able to locate those bodily and mental states of affairs (i.e. the diseases, impairments and defects) that tend to reduce our ability to realize our vital goals, we must certainly characterize them as meticulously as we can in physiological, neurological and biochemical terms. As a result we can then discover a number of diseases and other maladies without any direct communication with the bearer of the condition.

There are, thus, two crucial elements in the philosophy of health that I propose, which are not generally acknowledged. I insist that we should first select and characterize the concept of health (and not the negative counterparts such as disease, defect and impairment) as the most fundamental medical concept. From health the other concepts should be derived. And I propose that health should be defined in terms of a person's abilities, more precisely the person's ability to realize his or her vital goals.

I have, however, also written about topics in the philosophy of medicine where action theory is not immediately relevant. One such area is medical causation (1981), in particular causation of death (1983). These analyses of mine are little known and have not brought about much academic reaction. A contributory reason may be that my work in this area (although in English) had Swedish publishers and was not at all marketed outside Sweden. On the other hand, I would concede that my contribution to this area is only marginally original. I emphasized the complicated nature of the notion of cause and I attempted in particular to analyse and categorize various interpretations of the locutions: "x is a more important cause of z than y is" or "x is the principal cause of z". I applied this model of analysis to the problem area of causation of death, where we need to identify the "underlying cause of death". The underlying cause is supposed to be the principal cause, or the most important cause, of death in one of the possible senses of this concept.

In particular in my later projects I have been involved in ethics proper. The specific topics have been: the goals of medicine (for instance in Hanson and Callahan, 1999), animal welfare (2006) and, in particular, the nature of dignity and its place in health care (2004 and 2009). The latter project—which is perhaps the most original one—was placed in a European context, being a part of a broad programme supported by the European Commission. The programme was largely empirical and its aim was to investigate what older Europeans thought about the concept of dignity and in particular what constitutes dignity in health care and the care of the elderly. During the course of these studies I proposed a matrix of four types of dignity: *Menschenwürde*, dignity of merit, dignity of moral stature and dignity of identity. By distinguishing between and exemplifying various types of dignity I may have broadened the horizon somewhat in an ethical debate that has almost completely focused on the unique human value, *Menschenwürde*. My suggestions in this regard have caused certain reactions. A few of them demonstrate that my analysis ought to be further developed. Some such developments appear in my book *Dignity in Care for Older People* (2009). See below a brief discussion under heading 4.

Perhaps I should comment on my relation to normative ethics. I am basically a theoretical philosopher, trained within metaphysics, philosophy of science and philosophy of the humanities. This is quite an unusual situation for a philosopher of medicine, even more unusual for a philosopher of medicine with a humanistic inclination. This has meant two things in my case. My primary research focus has not been within ethics. My study of basic medical concepts, such as health, illness, disability and quality of life, has never been directly prompted by ethical questions. My interest has been scientific. I have asked questions such as: What is the content of these concepts which are so crucial for the science of medicine? It is another matter that, in the course of my analysis, I have come to the conclusion that all these concepts are basically normative, or at least normatively selected. This has shown me that the concepts are rooted in a deeply ethical enterprise, viz. that of getting rid of these evils that so frequently haunt human beings.

3. What, if any, practical and/or social-political obligations follow from studying medicine from a philosophical point of view?

I think philosophy can contribute greatly to medicine and health care also in many social and in general practical respects. This does not hold only for medical ethics, where the practical implications are obvious. It clearly also holds for theoretical analyses of medical science and of the clinical enterprise.

The outlook on medicine must be very different if one understands health and disease as a purely scientific affair without any (conceptual) connection to human suffering and disability. The whole enterprise risks under such circumstances becoming distanced and neutral. The patient would then, in principle, not have any say with regard to his or her health. The determination of diseasehood and illness would then be an entirely scientifically medical affair.

If on the other hand the primary perspective is the patient's and the patient's phenomenological illness, then the clinical situation must be entirely different. The patient's point of view becomes central and the biological knowledge would rather serve as an aid. The biological knowledge certainly broadens and deepens the understanding of subjective illness. Moreover, it can say much about the prognosis of certain bodily states. As I have argued above—and several times elsewhere—the biological knowledge is indispensable. What I wish to emphasize is that the starting point of the typical clinical situation is the appearance of subjective illness and the goal of the enterprise is the eradication of the illness and the creation of health. The goal is not per se to restore the body or mind of a person just to a (statistically) "normal" state of affairs.

An action-theoretic concept of health and illness has a particular importance for the enterprise of rehabilitation. If a person's health is understood as this person's ability to realize his or her vital goals, then rehabilitation must to a considerable extent become individualized. People have at least partly different goals. The differences between, for instance, a pianist, a carpenter and a professor are at least with regard to their professions considerable. For the pianist it is vital to be able to use the fingers in a virtuoso way. This is a crucial ability but not to the same extent for the carpenter. For the latter, on the other hand, it is of great importance to use the arms, both for lifting and also for hammering and fitting material. The professor, finally, can get on pretty well without functioning limbs, as long as there are means available for giving talks and for producing texts.

Conceptual analysis can have an eye-opening effect in several other areas of health care research. One salient example is the concept of need, which is frequently referred to in the context of medicine and health care. The Swedish Health Care Act says in one of its portal paragraphs that "health care should be provided according to the individual's needs." The Governmental Report on Priorities in health care from 1995 put a lot of emphasis on the patient's needs in making proposals for the prioritization of health care measures in our country. But what is a need of health care? And what is in general a need?

My colleague Per-Erik Liss (1993) and myself (1995, 2006) have spent a lot of time trying to demystify the notion of need. What we have claimed is that there is at bottom a bare and completely neutral concept of need, which could be analysed as: necessary condition for realizing a goal. Thus, anything is a need which is a necessary condition for realizing a particular goal that has been set by someone. The burden of analysis is then shifted to the goal. Is the goal self-evident? Who has set it? Has it been agreed upon?

Having said this, which is crucial enough, we must concede that there is a biological and psychological discussion on needs where the needs are often looked upon as absolute and not relative to any particular goals. Sometimes such needs are called basic needs. Liss and I wish to argue, however, that this mode of speech should be scrutinized from the point of view of the general concept of need. The following legitimate questions can be asked. What is a basic need? What distinguishes a basic need from other needs? Frequently the answer to the latter question is: the fulfilment of a basic need is necessary for the survival of the person or the organism. But then an answer has been given according to the above paradigm. The goal of a basic need is survival. Such a goal is clearly legitimate and almost universally accepted. But do we always refer to survival when we talk about basic needs? What about health and quality of life as goals of basic needs? We easily see that there is room for choice here. The purported biological notion of need is then revealed to contain a considerable element of ideology.

4. What do you see as the most interesting criticism against your own position in philosophy of medicine?

A. The risk of too expansive a concept of health.

It is frequently claimed that a holistic concept of health such as mine risks being too inclusive and vague. Or to be more precise, the most acute risk is that the converse concept, ill health, will include too much. If every person who is somewhat disabled in relation to his or her vital goals turns out to be ill, according to my conception, then we might come up with too many sick-listed persons. My concept would then turn out to be unreasonable and of no practical use in health care.

An interesting version of this criticism has been put forward by Schramme (2007). He formulates the following crucial purpose for a fruitful theory of health: it should be able to function as a gatekeeper against medicalization. Schramme claims that Boorse's biostatistical theory of health fulfils this purpose whereas mine doesn't. As an example Schramme says the following:

> Consider Lily, an athlete, who struggles, for her whole adult life so far, to become an accomplished high-jumper, but does not succeed... we may agree that it is one of Lily's vital goals to succeed, because not to succeed means that she is not minimally happy but actually quite angry and sad... Lily is unable to realize at least one vital goal, therefore she is not healthy according to Nordenfelt's definition. (p.14)

Schramme says that we certainly would not call Lily unhealthy in this case. We would only do that if she really suffered from a disease. The consequence of Nordenfelt's theory is thus counterintuitive.

I have already in a reply to my critics (2007) given an answer to this argument and I wish now to develop it. I say first that Schramme presupposes that I consider the concepts of health and illness to be contradictory in the strong sense that wherever we do not have complete health we have some degree of illness. In fact, for me health is a dimension ranging from a state of complete health to a state of complete illness. So, when Lily does not achieve what she has striven for so intensely – and this goal qualifies as a vital goal in my sense – the assessment is not automatically, I claim, that she is ill. The result is only that Lily's health is somewhat reduced. Her health is probably in general very good since she is capable of achieving most or all of her other vital goals.

I insist that it is reasonable to claim that her health is somewhat reduced, but that does not necessarily entail that there is reason for her to seek ordinary health care. The "cure" of her reduced health might instead be that she attempts to set more realistic goals for herself.

My answer to Schramme also contains a rebuttal of the biostatistical theory as an adequate one in this regard. I claimed that this theory risks including too little. If we require that there is a disease in the bio-statistical sense (i.e. a bodily state which causes a subnormal function in relation to survival) in a person in order to characterize him or her as ill, then one can wonder about the future for many people who are now normally (and I think legitimately) sick-listed. What should we do about people who are burnt-out or depressed and cannot perform their jobs or even take care of themselves? In many of these cases we cannot detect any biological malfunction in the Boorse/Schramme sense. Should we say that these people do not come within the scope of medicine or psychiatry?

Let me expand on this argument using a more conventional stance. Assume that we are not presupposing the Boorsian definition of disease but instead presuppose the conventional list of diseases and related conditions in the ICD classification. Then, Schramme might say, we can find among these diseases ones which cover the obvious cases of disablement shown in the burnt-out syndrome. We can there find symptom diseases such as Blackout, Fainting, Insomnia, Lethargy and Drowsiness. Yes, the classification in fact contains many negative symptoms, in fact many of those that we find in such "clear" cases of illness where there is no salient bodily dysfunction. But this will not help the Boorse/Schramme argument much, since their theory would no longer function as the firm gate-keeper against medicalization which Schramme wants it to be. How many instances of black-out or how much of insomnia and lethargy can count as legitimate illness on this account?

B. Why just disability? Why not suffering?

A common criticism of my concept of health and illness has to do with its uni-dimensionality. I emphasize the element of disability in ill health and say little (in the definition nothing) about suffering. I do so in spite of my acknowledging the fact that suffering is an extremely dominant factor in the phenomenology of illness.

My answer to this has been formulated in the following way (2001, p. 81). When our purpose is to define illness and we attempt to find as general a characterization of the concept as possible, then suffering cannot do on its own. That all illness entails suffering is not true. A person in a coma is ill but yet does not suffer. A person in a manic episode is ill but yet not suffering. Other cases of illness typically express themselves directly as disability, not necessarily entailing suffering; this holds particularly for the ones which are the results of defect and injury, such as paralysis and deafness. That people may suffer when reflecting upon their state of illness is a completely different matter. This suffering is not part of their illness. It is to be compared and equated with the grief that we experience when reflecting upon other sad events in our lives.

My purpose in proposing a definition of health and illness was to find a minimally sufficient way of characterizing the two concepts. The notion of suffering was not adequate. The notion of disability was more promising. This was so for two major reasons. First, disability covers many cases that suffering does not. This is obvious in the case of coma, deafness and paralysis. But also mania entails some disability, namely a disability in handling one's social life. Second, all suffering which is not trivial leads to some disability. Some theorists would say that suffering conceptually entails some disability. I think this is true. When a person is in great pain, as in an attack of migraine, that person cannot concentrate; thus, the possibility of doing any serious work, for instance, has disappeared.

In spite of this reasoning, I have declared that I am willing to reconsider the place of feelings with regard to health and illness. The feeling element is so conspicuous and plays such a role in the identification of most paradigmatic illnesses that it should perhaps have a more prominent place in the defining characterization of illness. A way of giving it such a place has been devised by Tengland (2007) where he proposes a disjunctive characterization of illness, using both the notions of disability and suffering.

C. But do vital goals have anything to do with medicine? The need for operationalization.

There are still a number of legitimate questions to be put with regard to my suggested theory of the concepts of health and illness. What exactly do I mean by the expression "vital goal"? I arrived at this notion after a lengthy analysis and dismissal of alternative choices (wants, needs etc.) All those alternatives are completely and saliently inadequate. My theoretical analysis is the following: A person A's vital goals are the states of affairs which are necessary conditions for A's long term minimal happiness.

I will not rehearse here my further explications of this notion. See for this my discussions in 1995 and 2001. My issue here is a practical one. How will we ever know which are people's vital goals? I think we must here distinguish between abstract concept analysis and the practical application of a concept. The expression "vital goal" belongs to the abstract discourse of definition. When we apply this concept in a particular situation we will certainly have to make operationalizations for various purposes. In the case of the care of a certain individual we will have to formulate the concrete goals that are of the utmost importance for this individual (for instance being able to run a household and being able to do a particular job). And in the case of more large-scale policy decisions for health care or health promotion we will have to find out what must be essential for most people in terms of natural resources, culture, technology and professional structure in the society in question. Health care and to some extent the operationalization of the concept of health itself will therefore be culturally dependent. But such operationalizations should, I argue, be made in the light of, and be informed by, my abstract definition.

The exact line between health and ill health is therefore impossible to draw from a conceptual analysis only. That is why we must for various purposes stipulate our use of the concept. We may for the purposes of health care in a particular clinic stipulate a bottom line that people declared to be healthy must have reached. Such stipulations, which are sometimes made in numerical terms, may for obvious reasons change over time and from place to place.

D. On the notion of dignity

In (2004) and (2009) I present, in the context of the care of the elderly, a typology of four kinds of human dignity: 1) *Menschenwürde* that pertains to all human beings to the same extent and cannot be lost as long as the persons exist; 2) The dignity of merit that depends essentially on social rank and formal positions in life. There are many species of this kind of dignity and it is very unevenly distributed among human beings. The dignity of merit exists in degrees and it can come and go; 3) The dignity of moral stature that is the result of the moral deeds of the subject; likewise it can be reduced or lost through his or her immoral deeds. This kind of dignity is tied to the idea of a dignified character and of dignity as a virtue. The dignity of moral stature is a dignity of degree and it is also unevenly distributed among humans; 4) The dignity of identity that is tied to the integrity of the subject's body and mind and to the autonomy of the individual, and is in many instances, although not always, also dependent on the subject's self-image. This dignity can come and go as a result of the deeds of fellow human beings and also as a result of changes in the subject's body and mind.

This typology has been challenged in certain respects, for instance by Wainwright and Gallagher (2008). Their main complaints concern the place and importance of this typology in elderly care. They doubt that all four types can "form the basis of human worth". They find, in particular, that the category of dignity of merit is irrelevant in the context of care.

They also find my characterization of the dignity of identity to be insufficient. It seems fatal, they think, to tie the dignity of identity so closely to the integrity and autonomy of the person. An individual can maintain self-respect and dignity of identity even when his or her integrity or autonomy has been violated.

Space prevents me from commenting on details in this criticism. I wish here, however, to make a few statements in order to avoid certain misunderstandings. First, although I present this kind of typology in the context of the care of the elderly, my principal aim has been to make a basic and general analysis of the concept(s) of dignity. The fact that I mention the four types of dignity parallel to each other does not mean that I find them all to be equally relevant in health care or the care of the elderly. It is clear from my presentations, and it is particularly developed in my 2009 book, that *Menschenwürde* and dignity of identity are the two types that are most relevant in the context of care. The dignity of moral stature is also relevant, but in a different way than the others. We should not require a particular moral status on the part of the object of care, viz. the patient or the client, in order to provide adequate care. We should however require a high moral status on the part of the provider of the care. The provider should act morally, i.e. in a dignified way. Second, my idea in providing this list is not to say that these kinds of dignity can be added to each other in order to sustain the idea of human value in general. Instead, as I see it, they represent quite different kinds of value that cannot be added to each other. The most important value is the general *Menschenwürde* and that is what we normally call the specifically human value. Third, the critics are quite right in observing that there are relations between the items in my typology. To trace these relations might be a project worth pursuing. I have myself initiated such an enterprise. I have observed that the dignity of moral stature can be construed as a special case of dignity of merit. I have compared *Menschenwürde* and dignity of identity and noted resemblances and differences.

My fourth and final comment is related to an interesting observation by Wainwright and Gallagher. They are sceptical with regard to my way of connecting dignity of identity to integrity and autonomy. They say that an old person may very well become distorted by illness and other bodily changes. His or her autonomy may have become diminished. Still, the person may have retained his or her self-respect and may therefore, they claim, have retained a high degree of dignity of identity.

A full reply to this crucial observation requires much space and much reflection. I will now just make a start. First, the comment by Wainwright and Gallagher points to the difficult relationship between a person's self-respect (or self-image) and his or her dignity of identity. As Wainwright and Gallagher claim, one person with a violated integrity can retain self-respect and a strong self-image whereas another person with an equally violated integrity can be shattered and feel that he or she has "lost the dignity". But does it follow from this that their dignities of identity are different?

According to my understanding the person's self-respect or self-image influences his or her dignity of identity but does not completely determine it. If it did we would have to draw some counterintuitive conclusions. Assume that a mentally retarded boy is grossly humiliated by another person who keeps ridiculing him in front of others. The young boy, however, does not understand the insolence. He does not feel disturbed or feel that his integrity has been violated. On the other hand, many bystanders are deeply disturbed and they would, I claim, rightly insist that the person's integrity has been violated and that his dignity of identity has been reduced.

Thus although I see the force of Wainwright and Gallagher's observation it need not undermine my basic understanding of the concept of dignity of identity.

5. With respect to present and future inquiry, how can the most important problems concerning Philosophy of Medicine be identified and explored?

A. The goals of medicine.

One of the most crucial and profound issues in the philosophy of medicine of today and tomorrow concerns the goals of medicine and health care. What should be the concerns of medicine in the future and what should fall outside the scope of medicine?

These issues are currently being discussed in the international literature. One of the most crucial contributions is the one by Mark Hanson and Daniel Callahan: The Goals of Medicine: The Forgotten Issues in Health Care Reform (1999). Here some of the traditional goals of medicine are scrutinized and some more sustainable goals are proposed. Among the traditional goals are the following: the saving and extending of life; the promotion and maintenance of health; and the relief of pain and suffering. All of them entail problems. What does the goal of saving and extending life mean when machines can sustain the bodies of those who would, in earlier times, have died? What should promotion and maintenance of health mean in the case of those people who have reached the age of 100? What should relief of pain and suffering mean when we talk about the anxieties of daily living and the psychological and existential problems that people face in ordinary life?

Hanson and Callahan themselves propose a number of carefully argued goals (which have been formulated within an international research programme called The Goals of Medicine). Most of them are similar to the traditional ones but considerably qualified: the prevention of disease and injury and the promotion and maintenance of health; the relief of pain and suffering caused by maladies; the care and cure of those with a malady and the care of those who cannot be cured; and finally, the avoidance of premature death and the pursuit of peaceful death.

Hanson and Callahan's proposals entail a considerable step forward but much more has to be said, not least in terms of further specifications and delineations. For instance, how should we distinguish a malady from an existential problem? Or, should we make such a distinction? To what extent is existential care a medical problem and to what extent is it a pastoral one? To what extent should the ailments of the elderly be characterized as maladies?

Other authors note that medicine is now in a turmoil where economic and other societal concerns become directly involved in the definition of medical problems. George Khushf (2007) claims, for instance, that implicit economic considerations are now being made explicit and directly incorporated into standards of medical care. Moreover, there is an overlap between administrators and clinicians. In many ways, we are, says Khushf, moving from an individually based to an organizationally based practice, and from an individually oriented clinical encounter to one that simultaneously addresses individuals and populations. "In the United States, where many of these developments have been quite aggressive, we find a deep challenge to the classical jurisdictions of clinicians and administrators" (Khushf 2007, p. 25). Many traditionally educated doctors and other medical staff see this development as dangerous or even as a distortion of medicine.

B. Future health enhancement and the future doctor

I cannot here address all these profound issues that face the medicine of tomorrow. I will here confine myself to commenting on the future role of the doctor. I shall pose the following questions. What should be the role of the doctor in future health care and how should he or she be educated? Should the doctor – as is still essentially the received view – be preoccupied with the diagnosis and cure of salient diseases? Or should he or she be more generally involved in the pursuit of and protection of positive health, where the cure of disease is only a partial affair? Or do we need a new kind of professional, a health generalist, to reflect and act with regard to these over-arching matters?

I will here contribute to this discussion by drawing some conclusions from my own health analysis. When one looks more closely at my suggested concept of health one can see that it involves in fact a three-place relation. The first term of this relation is the agent's ability, the second is the agent's vital goals, and the third is the standard circumstances surrounding the agent. Another way of expressing this is to say that health is a relation between the agent, his or her vital goals and the environment. Moreover, it is a relation of an equilibrium kind, where the abilities are supposed to match the vital goals in the standard environment.

An immediate consequence of this is that work for health ultimately must be work to establish the particular relation that should hold between the agent, the goals and the environment. Such work can indeed be achieved via traditional medical care, i.e. the treatment and cure of diseases, injuries and defects. One can go directly via the body or mind of the person who is ill. Such is also our normal understanding of medicine.

But it is clear that if something is a relation of the equilibrium type that we envisage in the case of health, then this equilibrium can come about through other measures. Goals can be lowered so as to match a person's set of abilities, and circumstances can be changed so as to give a person better opportunities to execute his or her abilities to realize the goals.

In principle, then, in addition to the agent care, which is the typical kind of care that exists in medicine and health care in general, we must also conceive of what could be called goal care and environment care. The former would be a care that is directed towards a person's goal-formation, for instance concerning the realism in this person's goals in life. The latter form of care would be geared towards changes in the person's standard environment. (Observe that this presentation is highly simplified. Goal care as a part of health care cannot deal with any kind of goals. We must have in mind that health in my understanding is related to vital goals which form a particular subclass of goals.)

The question now is: to what extent does traditional medicine deal with all three kinds of care and to what extent does traditional medicine have the competence to deal with all three kinds of care? Should we view medicine as only one kind of care, that which is directed at the treatment and cure of the biological (and sometimes the psychological) basis of a person's ability? Or, should we broaden the basis and competence of medicine? And what repercussions would this have on medical education?

Let me now analyse a concrete case. Consider a woman in a Western society who seeks medical help for a number of problems. She is unusually anxious, she is sometimes depressed and she also has some somatic symptoms in the form of headache and pain in her stomach.

She comes to a doctor with a typical Western medical training, a training that contains 95% natural science or somatic clinical medicine. The doctor in question is a very conscientious person, makes a very thorough physical examination of the woman, but discovers nothing pathological in the traditional sense.

It is natural for the doctor with his or her training, to try to solve a problem (which at least superficially has to do with a person's body or mind) within a disease-oriented framework. But perhaps the problem, although a problem of health, is not of this kind. There may be another set of questions to be asked and another set of solutions to be tested. A careful analysis of the woman's situation may cause the following story to emerge. The woman fears that she is in the process of losing her whole family and she does not understand why. Her husband has told her that he is leaving her. She reacts in desperation and frustration and becomes unable to go to work.

It seems clear that this woman does not require standard medical care, although her health is not good. In order to alleviate her suffering and disability one must instead try to improve her situation. Her ill health is very clearly dependent on the environment, or at least her perception of the environment. In all likelihood both her mental suffering and her somatic ailments will disappear if her situation improves.

We may find further examples by looking systematically at the concept of health that I have suggested. In our previous example the ill health was dependent on the circumstances. But it is easy to see that it can also be dependent upon the person's set of vital goals. A young extremely ambitious boy may aspire to become the Olympic champion in high jump. He puts all his energy into this enterprise. He neglects school and family. However, he does not at all get the results that he had expected. The chances of his realizing his ambitions are remote. As a result he soon becomes highly stressed and unhappy. He refuses, however, to understand that his goals are hopeless and he keeps training. Finally, the boy ends up in a psychiatric clinic in an acute crisis and depression. This boy is evidently unhealthy. But there is no particular disease, whether somatic or mental, which accounts for this state of ill health.

It is probably true that the clinic when it works at its best can handle several of the variants of ill health that my triangle of health indicates. But the question is whether this is done because of or in spite of the prevalent disease-oriented philosophy of medicine.

There is at the very least some conflict here. There are two broad concepts of health operating beside each other. One concept of health, the more narrow one, forms the basis of the medical education of doctors and as a consequence forms the medical perception of doctors. At the same time the World Health Organization works within a holistic health framework, and some medical philosophers, some sociologists and most theoreticians within the paramedical disciplines propose a holistic understanding of health.

It is not clear that the medical profession, as its representatives are at present trained, has been afforded a salient task and has been afforded all the intellectual resources for such a task. The doctors, to take them as the primary example here, are trained to deal with one — albeit an extremely important and the technically most complicated — part of health care. But they have in general no particular education concerning the other parts. What is the answer to this dilemma?

Allow me here to speculate a little. I think that there are two principal ways to handle these issues. One is to base a reform on the already existing division of labour. Health care is indeed not exhausted by the work of doctors. There are a multitude of paramedical staff, nurses, physiotherapists, occupational therapists, psychologists and social workers. And indeed all these people do in practice perform much of the humanistic work required, both in terms of diagnosis and in terms of cure and rehabilitation. There are several well-functioning health care teams, including all these different kinds of staff, all over the world. I am not denying this fact.

The problem, however, is that in our present system there is a risk of one paradigm remaining the dominant one. The teams that we have at present are almost always — although there are interesting exceptions — headed by a physician or a psychiatrist. There is a built-in hierarchy in most Western health care. The bearer of the old medical paradigm is almost always the person who monitors and decides how the clinical work should be performed. Given the analysis that I sketch here, it is not clear that this ought to be the case. It seems as if a more reasonable division of labour ought to entail a communication between equals, where the reasoned discussion between them should be the platform for the health care to be chosen in a particular case. But in order to realize such a reasoned discussion between equals, it is of course necessary that the paramedical professions receive a status which is equal to that of the doctor. This presupposes probably an extended education on their part, entailing some deep study in social sciences and humanities to supplement their present education.

I have a second option which may sound more unrealistic but which I wish to mention in order to open our minds to new alternatives. I think that one could consider the creation of a new profession and thereby a new member of the health care team, namely a health generalist. This would be a person whose speciality should be the analysis of health problems and who should not have any particular bias in the search for solutions to such problems. He or she should be open to the search both for organic causes and for the existence of existential problems and intolerable environments. The health generalist should, then, be a very qualified person with just as much traditional medical education as education in non-medical health matters.

The task of the health generalist could be to serve as a kind of gate-keeper (without its negative restrictive connotations) who sincerely analyses the patient's or client's problems. This person should see to it that a person's problems are correctly identified so that a purported somatic problem that is situationally caused or has to do with goal setting receives its proper treatment and care. Conversely, a deep somatic disease can for a long time be masked as an existential problem and has in its turn to be properly identified.

References

Texts by Nordenfelt referred to:

Causation: An Essay (1981) Akademilitteratur, Stockholm.

Causes of Death (1983) Forskningsrådsnämnden, Stockholm.

On the Nature of Health (1987/1995) D. Reidel Publishing Company, Dordrecht.

Action, Ability and Health: Essays in the Philosophy of Action and Welfare (2000) Kluwer Academic Publishers, Dordrecht.

Health, Science and Ordinary Language (2001) Rodopi Publishers, Amsterdam.

The varieties of dignity, Health Care Analysis (2004), 12, 69-81.

Animal and Human Health and Welfare: A Philosophical Comparison (2006) CABI, Wallingford.

The concepts of health and illness revisited, Medicine, Health Care and Philosophy (2007), 10, 5-10.

Establishing a middle-range position in the theory of health: A reply to my critics, Medicine, Health Care and Philosophy (2007), 10, 29-32.

Rationality and Compulsion: Applying Action Theory to Psychiatry (2007) Oxford University Press, Oxford.

The Concept of Work Ability (2008) P.I.E. Peter Lang, Brussels.

Dignity in Care for Older People ed. (2009) Blackwell-Wiley, Oxford

Other references:

Boorse, C. (1975) On the distinction between disease and illness, Philosophy and Public Affairs, 5, 49-68.

Boorse, C. (1977) Health as a theoretical concept, Philosophy of Science, 44, 542-573.

Boorse, C (1997) A rebuttal on health. In: What is Disease? Biomedical Ethics Reviews, eds. J. Humber and R. Almeder, Humana Press, Totowa, NJ.

Canguilhem, G (1978) On the Normal and the Pathological, D. Reidel Publishing Company, Dordrecht.

Fulford, K. W. M. (1989) Moral Theory and Medical Practice, Cambridge University Press, Cambridge.

Hanson M. and Callahan D. (1999) The Goals of Medicine: The Forgotten Issues in Health Care Reform, Georgetown University Press, Washington DC.

ICIDH (International Classification of Impairments, Disabilities and Handicaps) (1980) Publication of the World Health Organisation, Geneva.

Liss P-E. (1993) Health Care Need: Meaning and Measurement, Avebury, Aldershot.

Pörn I (1993) Health and adaptedness, Theoretical Medicine, 14, 295-303.

Priorities in Health Care: Ethics, Economy, Implementation (1995) Final report by the Swedish Parliamentary Commission 1995:5. The Ministry of Health and Social Affairs, Stockholm.

Schramme T. (2007) A qualified defence of a naturalist theory of health, Medicine, Health Care and Philosophy, 10, 11-17

Khushf G. (2007) An agenda for future debate on concepts of health and disease, Medicine Health Care and Philosophy, 10, 19-27.

Tengland (2007) A two-dimensional theory of health, Theoretical Medicine and Bioethics, 28, 257-284.

Wainwright, P. & Gallagher, A. (2008) On different types of dignity in nursing care: a critique of Nordenfelt, Nursing Philosophy, 9, 46-54.

13

Onora O'Neill

Professor Emeritus of Philosophy

University of Cambridge

---

1. Why were you initially drawn to Philosophy of Medicine?

Like most people trained in philosophy in the 1960s I was not initially drawn to the philosophy of medicine, to medical ethics, or indeed to other areas of 'applied' ethics. When I started writing philosophy my interests were in reasons for action and in normativity, and many of the articles that I published were about topics such as principles and act descriptions, abstraction and idealisation, obligations and rights; I have also worked on many aspects of Kant's philosophy throughout my professional life. My work shifted in the 70s, when political events led me to think about a range of more practical questions. I started to write more in political philosophy, and in particular on conceptions of equality and liberty, and on issues of justice, poverty and development. But still I barely looked at medical ethics.

This was not because I was unaware that other philosophers were beginning to work more on the philosophy of medicine and on medical ethics. I was working in New York for much of the 1970s and went to meetings of the *Society for Philosophy and Public Affairs* (which shared ancestry with the eponymous journal, but was distinct). In 1976 – I think—I was at a meeting of physicians and philosophers which discussed genetic modification. At that time the modification was of bacteria, and the ethical concern was about what is now called containment: was it wrong to sluice modified bacteria down the drain into the sewage system? I cannot remember what conclusion was reached, but was struck by the surprised comment of one elderly physician, who said that when he had trained there had only been three topics in medical ethics: confidentiality, referrals—and billing! I knew that others were trying to introduce more philosophical approaches into medical education, which would take the curriculum beyond confidentiality, referrals— and billing. I took no part. My work was in political philosophy and ethics, and on the philosophy of Immanuel Kant.

Then in the 90's I began to pay some attention to a limited range of questions in bioethics, initially in practical rather then philosophical contexts. I served on a range of committees: I was a founder member of the *Nuffield Council on Bioethics*, and chaired it for a time—and have remained close to it because I subsequently chaired the *Nuffield Foundation*, which founded and part funds the Council. I was a member of various public bodies, including the *Human Genetics Advisory Commission*. I wrote very little on medical ethics, or more broadly on bioethics, during these years, except by helping to draft reports. The reports I contributed to dealt not with clinical ethics but with topics such as uses of genetic data, genetics and insurance and the use of human tissues. In the 90's I published a few papers on these non-clinical issues, in particular on genetics and insurance. I still did not think of medical ethics as an area in which I had much to say. In 2000 I was nearly 60, and although I had published a lot in philosophy I could point only to two short, incidental papers in medical ethics.

2. What does your work reveal about Philosophy of Medicine that other academics, citizens typically fail to appreciate?

A late starter cannot expect to contribute much to the heartland of medical ethics, but can perhaps be helpful by bringing more distanced perspectives to certain topics. I think I have been able to contribute such perspectives in two areas.

The first contribution took shape when I realised that much of medical ethics was preoccupied with conceptions of informed consent, and by claims that informed consent was required in order to support or protect 'patient autonomy'. This concern with autonomy was often misrepresented as Kantian. Because I had worked a lot on Kant, I knew that this claim was false: Kant thought of autonomy as a matter of acting on principles that can be principles for all, and argues that Kantianly autonomous action is central to morality. Although individuals need capacities for free choice if they are to adopt principles that would count as Kantianly autonomous, they often use those capacities to choose in ways that are not Kantianly autonomous. Once we focus on one or another conception of individual autonomy, the links that Kant drew between morality and his conception of autonomy do not hold, and any connection between morality and individual autonomy remains to be established. This proved harder than some imagined.

The problem was greatly exacerbated because a plethora of conceptions of individual autonomy were in circulation, not only in medical ethics, but in other parts of ethics and political philosophy during the 80s and 90s—and indeed to this day. Some saw all choices as autonomous; others saw choices as autonomous only if they were informed, or reasoned, or reflective in specific ways, that were often highly demanding for patients and practitioners, indeed sometimes impossible. There was, it seemed to me, little prospect of justifying informed consent requirements in medical or research practice simply by claiming that they secured (some conception of) individual autonomy. Some of these claims were pointless exaggerations that invoked requirements that consent be 'fully' or highly informed, or 'fully' or highly explicit: these standards set impossible hurdles for patients and research subjects, and uncertain standards for clinicians, researchers and anyone using patient data. Other accounts of informed consent were less exorbitant— occasionally even feasible for all parties— but their justification did not rest securely on any one conception of individual autonomy. In various articles and books written between 2002 and 2007, and in particular in *Rethinking Informed Consent in Bioethics*, written jointly with Neil Manson, I argued that there were other and more plausible ways of justifying consent procedures that did not appeal to any of the variegated conceptions of individual autonomy. I came to think that consent, of a pedestrian but achievable sort, matters not because it secures some prized version of individual autonomy, but because it provides adequately robust assurance that certain basic obligations— such as obligations not to coerce, deceive, manipulate or do violence to patients or research subjects—are not breached. In consenting we waive others' obligations not to breach these standards if they act for specific purposes: for example, in consenting to surgery we permit another to do something that would *otherwise* count as assault.

The second area in which I found a more distanced perspective helpful was in thinking about public health ethics. Once more my approach queried excessive individualism. I had already argued that placing excessive weight on conceptions of individual autonomy, as well as appeals to supposed 'rights to health' for individuals, are hard to interpret and even harder to justify, despite their popularity. I extended these lines of thought to the case of public health. In thinking about population health it is important to take account of harms and benefits that affect many, whether the interventions that produce them are targeted on identifiable individuals (e.g. immunisation) or not (e.g. safety standards). The world we live in transmits and disperses benefits and harms among agents in ways that are not wholly foreseeable or separable, and we cannot seek (let alone obtain!) consent for everything that happens to others, or even for everything that happens to them as a result of our action. Public health measures, I argued, are better viewed through the lens of political philosophy, which has always had to take seriously the reality that consent cannot be sought for all action. Medical ethics and philosophy of medicine need to set aside any exclusive focus on individuals: this may be appropriate for large parts of clinical ethics, but even clinical practice requires many provisions that have to be uniform for populations, so cannot be held to require consent (e.g. standards for clinical training; ways of determining the safety and efficacy of drugs). In shifting focus from ethics to political philosophy, it becomes clear that arguments that bear on equalities and poverty, on justice and development, also bear on public health.

3. What, if any, practical and/or social-political obligations follow from studying medicine from a philosophical point of view?

I have little to say about this, perhaps partly because I do not really see myself as studying medicine from a philosophical point of view. I have done no more than study certain philosophical and policy arguments that bear on some parts of medical practice, and on some related areas of scientific research, practice and communication; and that very selectively. Insofar as I have taken on practical and other tasks in those areas, it has been largely by contributing to public reports and policy debate.

Since around 2000, much of the policy work that I have done in this area has been undertaken by way of membership of and contribution to the work of a series of Select Committees of the UK House of Lords—in that context I have found myself working successively on the regulation of IVF and stem cell research; on prospects for genomic medicine; on nanotechnology and food, and on behavioural change. In other contexts I have contributed to a range of reports on the use of human tissues, on uses of genetic information, on conceptions of privacy and data protection; on patient safety; on professional and research ethics and on the implications of making assisted dying lawful. I have also served as a Trustee of several charities with commitments in these areas; some of which seek to improve and enable better communication of science, including medical research, to the wider public. Insofar as I have found philosophical experience useful for this work, it is usually because it helps in formulating and drafting normative and other arguments, and not because I have acquired specific expertise in medical ethics.

I do this work not because I think that anyone who works on medical ethics or studies medicine from a philosophical point of view thereby incurs specific political or social obligations, but rather as an intrinsically interesting, if small, bit of public service.

4. What do you see as the most interesting criticism against your own position in philosophy of medicine?

My criticisms of appeals to autonomy in medical ethics led me to query the marginalisation of trust that has been common in Western medical ethics since the 1970s. Advocates of 'patient autonomy' frequently dismiss trust as naïve and risky (although many doctors and rather a lot of patients still think it necessary). It has been frustrating to work on arguments intended to support a more intelligent and more practical view of trust over nearly a decade, and to find that I often meet with initial agreement—which then turns out to have no sticking power.

In my view, the persistent hostility to trust in medical ethics typically arises because trust is seen as a *generic* attitude to *types* of cases ('Do you trust doctors?'), thereby bracketing everything that differentiates cases and erasing everything that is relevant to the practical task of placing and refusing trust intelligently. As I see it, trust need not be blind trust, and placing and refusing trust intelligently is an indispensable *practical* task in all human affairs, including medical practice. My critics persistently view trust as a generic attitude, and suggest that relying on trust is infantile and risky. As I see it, we need to recast discussions of trust, and focus less on the empirical studies of the attitudes people evince, and more on normative analysis of the judgments they make. In particular we need to think more about the ethical and epistemic norms that matter for the intelligent placing and refusal of trust.

I cannot pretend that my arguments have been widely accepted. I was lucky enough to reach a very large audience when I first presented them in the BBC Reith Lectures for 2002, published as *A Question of Trust*; in the same year I published 'Autonomy and Trust in Bioethics'. Since then I have been energetic in exploring my arguments and conclusions with a wide range of audiences. Yet, nearly a decade later, I find that many people say that they agree with a lot of the specific points that that I make about trust, but still think trust irrational and risky, and suggest that suitable forms of accountability offer us a more reliable, trust free way of organising medical practice and other parts of life. I think that the evidence suggests that systems of accountability often have great weaknesses—and that they always presuppose at least some forms of trust. I do not know whether or how far these disagreements can be resolved, but continue to argue the case.

5. With respect to present and future inquiry, how can the most important problems concerning Philosophy of Medicine be identified and explored?

I suspect that it is not possible to identify or even to list 'the most important problems concerning Philosophy of Medicine'. Philosophy, including philosophy of medicine, does not consist of a series of discrete problems that can be identified and picked off one by one. 'Problems' are interrelated, and many of them are persistent. Large areas of medical ethics are matters of perennial concern: the relations between doctors and patients, the tension between care and cure, ways of weighing the costs and benefits of technologies, and the proper balance between public and private funding are all perennial concerns rather than problems that can be resolved once and for all.

I suspect that medical ethics is sometimes thought of as consisting of a set of discrete 'problems' that can be 'identified and explored' *seriatim* only because diseases and treatments can be listed and may be seen as discrete *up to a point*. However, lists of diseases and treatments change, and often changes cannot be anticipated. The very distinctive problems raised by HIV/Aids or by SARS or by new variant CJD could not have been anticipated, identified or explored before their emergence. Nor could the very distinctive problems raised by new reproductive technologies from IVF to surrogacy, from pre-implantation diagnosis to reproductive cloning have been identified and explored before the relevant innovations were achieved—or seemed imminent.

So there are always good reasons for asking whether the most urgent ethical problems in medicine are actually being addressed. In my view, many of us working in ethics, as well as in philosophy of medicine, have failed for nearly three decades to take population ethics sufficiently seriously. An individualistic focus has led to excessive preoccupation with 'reproductive autonomy' and so with unwanted infertility in rich societies, but has marginalized consideration of persisting unwanted fertility in poorer societies. It is a bitter irony that the topic is regaining attention only because increasing human populations have contributed to widespread environmental harm.

13. Onora O'Neill

14
Peter Rossel

Unit of Medical Philosophy and Clinical Theory.

Department of Public Health, University of Copenhagen

---

1. Why were you initially drawn to Philosophy of Medicine?

When I finished school it seemed quite unlikely that I was going to have any sort of academic career. After school I got educated part of the time in my farther's bookstore, which I was supposed to take over in due time and afterwards I was drafted for the navy for $1^1/_2$ years. I spent the last year at the Nato Navy Headquarter in Kiel, Germany – it was in 1965-1966. This was during the Vietnam War and the staff included several Americans who had served as marines in that war. Being opposed to the war, I remember especially one reply: "Oh Peter, we Americans are just fighting for your freedom" – a reply that just increased by developing critical attitude. Realizing that my future was neither as the owner of a bookstore in a middle-size provincial town in northern Jutland nor in the navy I took studentereksamen (approximately a high school certificate) to get eligible to enrol at a university. Having been in close touch with the realities outside the academic world was, I later realized, a valuable and for me indispensible asset.

Then in 1969 I started at the University of Aarhus, Denmark to study History of Ideas – being seven years older that my co-students. The Department for the History of Ideas was in the beginning of the 1970ties dominated by continental trends in the form of different kinds of neo-marxist schools (not classical marxism), structuralist and post-structuralist thinking like e.g. Althusser and Foucault. One of the preoccupations was a critique of so-called reactionary and bourgeois science and especially positivism was seen as an expression and a legitimation of the status quo. The criticism of this "evil" positivism spurred my interest and I found that the term was used indiscriminately to cover Comte's positivism as well as logical positivism and Popper's critical rationalism. So I began to study logical positivism and especially Popper's philosophy of science, both of which I found intellectually liberating for their commitment to rationality and clarity of expression, as opposed to the often obscure language of continental thought.

My first paper was about The Lysenko Affair (1), which I presented at a conference in 1977 on "History of Science" at University of Aalborg, Denmark, with the main theme of the conference being the relationship between history of science and philosophy of science. Here I gave an analysis of the proceeding from a conference in U.S.S.R. held January 31st to August 7th 1948. This conference meant a destruction of genetic research as it was practised in the West and hitherto by the eminent plant geneticist N.I. Vavilov in the U.S.S.R. "Mendel-Morganism" was rejected as an idealistic, metaphysical, formalistic, reactionary, bourgeois pseudo-science in conflict with dialectical materialism. Instead Lysenko's theory of heredity was hailed as a materialistic, progressive, proletarian and dialectial science.

Another speaker at the conference was Stig Andur Pedersen (then University of Copenhagen) and the contact with him became all important for my turning to philosophy of medicine and especially medical ethics. Up to then my main interest had been general philosophy of science, history of logic, philosophy of biology and evolutionary theory, the new genetics and the emerging *ethical aspects* of the developments in genetics. Accordingly, the subjects of my thesis were the Popper-Kuhn controversy, the Lysenko Affair and history of formal logic and I had a vague idea of an academic career in the history of logic. But the contact with Stig Andur Pedersen changed all that in a decisive way.

The publisher Gads Forlag was going to publish a new series in philosophy (in Danish) with the first volume in 1978 and the second volume planned for 1979. At the suggestion from Stig Andur Pedersen to the editor the second volume could be about medical ethics, which he (rightly) saw as an emerging important subject within philosophy of medicine. Being in touch with him since the conference in 1977 and with his knowledge of my interests in the ethical aspects of the developments in the new genetics, he contacted me in 1978 with the proposal that I undertook the task of writing the book. Being without a job after just getting my degree from the university I applied (needles to say with support from Stig Andur Pedersen) for a grant from The Danish Research Council for the Humanities to do the necessary research for writing the first book from a philosophical perspective in Danish on medical ethics. Despite my interest in the ethical aspects of the new developments in human genetics my interest had by now shifted to the ethical aspects of the ordinary doctor-patient relationship and the ethics of research on human beings. Luckily I got the grant and the book "Medicinsk Etik" (2) was published in 1979.

In accordance with this shift of perspective some chapters in my book are historical analyses of the ethical aspects of the ordinary doctor-patient relationship as formulated in documents of the medical profession e.g. the Hippocratic Oath and Perceival's influential "Medical Ethics" from 1803. In short a focus on traditional medical paternalism. One of the most central chapters of the book, however, is a moral philosophical analysis of the proceedings of The Doctor's Trial and the formulation of the Nuremberg Code (1947). Of special interest for my analysis with respect to the ethics of human subject research is how the defendants tried to justify ethically research on human beings *without* their voluntary consent. And, especially interesting are the examinations and cross-examinations by the prosecution, the judges and the defence counsels of the defendants and the expert witnesses for the prosecution in the process that led to the formulation of the Nuremberg Code with its absolute requirement of informed consent.

The next chapter on The Declaration of Helsinki contains a critical moral philosophical analysis of the revised version of Declaration of Helsinki 1975 (DoH (1975)) – a chapter that turned out to be rather controversial. The Declaration of Helsinki is a central document and has been described as the "cornerstone" document pertaining to medical research and as "the most widely recognized source of ethical guidance for biomedical research" (3). The first version of the declaration was adopted by the World Medical Association in 1964 (DoH (1964)). In 1975 the declaration was extensively revised and although there were minor revisions in 1983, 1989 and 1996 "it is effectively the 1975 version which became the guiding document for the ethics of research involving human subjects for a quarter of a century" (3). The revised version was adopted by World Medical Assembly, Tokyo 1975. The drafts of DoH (1975) were formulated by a Scandinavian group of physicians: Clarence Blomquist, Sweden, Erik Enger, Oslo and Povl Riis, Denmark. As in the original Declaration of Helsinki (1964) a distinction between "therapeutic research" and "non-therapeutic research" structures the whole document and this distinction plays a crucial role as to the requirements of informed consent for theses different kinds of research. With respect to "therapeutic research" in the section "Medical research combined with professional care (clinical research)" one thus finds *"If the doctor considers it essential not to obtain informed consent, the specific reasons for this proposal should be stated in the experimental protocol for transmission to the independent committee.* (My emphasizing)"

This so-called "escape clause" is obviously problematic from an ethical point of view as it pertains to patients who are competent persons. The problematic character of the clause should be clear as it makes it possible to suspend the requirement of informed consent and involve patients in clinical research e.g. in a randomized clinical trial (RCT) without informing them that they participate as patients/research subjects in such a randomized study. The "escape clause" makes this possible and in Denmark it became a reality. A case-study of such a research project will be presented in my answer to the next question.

In the chapter I undertook a critical analysis of the "escape clause" within the structure of DoH (1975) and I raised the concern whether the clause might legitimate that the interests of science and society would prevail over the concern for the interests of the individual person. This is not a trivial point within the framework of the declaration as its Basic Principles states: "Concern for the interests of the subject must always prevail over the interests of science and society". (I.5)

As mentioned the chapter turned out to be rather controversial – at least among (some) physicians. The book was reviewed in Ugeskrift for Læger by Povl Riis, one of the authors behind the formulation of DoH (1975) and thus the "escape clause" I had problematized! It was a scathing review and I was so-to-say denied clinical as well as historical and philosophical qualifications. Okay clinical qualifications, yes, but as a very young historian of ideas I experienced myself excommunicated from the medical world – and without a job. Neither could I look forward to a position at The Department for the History of Ideas, given the dominant philosophical orientation at that institution.

Happily, again through a research grant from The Danish Research Council and with the hospitality of The Hastings Centre, New York, me and my family in August 1981 went to U.S.A. for a one year stay at the Centre. It proved to be one of the most wonderful years in our life, both because of the hospitality towards me and my family and the intellectual stimulation at The Hastings

Centre. Around at that time were Ron Bayer and Art Caplan at the staff and Kai Nielsen, University of Calgary, Canada as a senior visiting scholar. Kai Nielsen, a moral philosopher and philosopher of religion, who was now engaged in political philosophy and theories of justice and who had lately become an analytical marxist. It was Kai who introduced me to the method of wide reflective equilibrium as a coherence method of justification in ethics.

During my stay at The Hastings Centre I got to know that Carl Erik Mabeck had been appointed professor of general practice at the Faculty of Medicine, University of Aarhus. I had met Carl Erik Mabeck a couple of times before going to the States and I knew of his interest in medical ethics. So I contacted him to enquire whether it would be possible to work together on issues in medical ethics on my return from the States. I got a positive reply and back in Aarhus in 1982 I contacted Carl Erik. Again, after getting a research grant from The Danish Research Council for Medicine I got employed as a research fellow at the Department of General Practice. This position should prove to be my way into medical circles and I will forever be grateful to Carl Erik for his help and support.

My research project was about the ethical aspects of the decision making process in general practice with special focus on information and consent in daily clinical situation. The project fell in the domain of *descriptive or empirical ethics* where I by means questionnaires containing vignettes of clinical situations in general practice and in depth interviews undertook a study of moral conceptions among general practitioners and lay people. The construction of questionnaires and the interpretation of interviews were based on certain moral philosophical assumption such as the distinction within normative ethics between consequentialist and deontological reasons, as well as concepts of autonomy, a principle of respect for autonomy and paternalism. One important finding was that a reason for not acting paternalistically among general practitioners was *not* respect for patient autonomy, but because the paternalistic act in question would violate *an agent-relative deontological constraint* against lying for or deceiving the patient (4).

My next and decisive step into the medical community came when the then Dean of The Medical Faculty, Kjeld Møllgård, University of Copenhagen in December 1987 called me to ask whether I would move to Copenhagen to start up a new obligatory course in Medicinsk Videnskabsteori (Theory of Medical Science) for medical students. The course had been introduced through a revision of the curriculum and had to start up by February 1988. A forerunner of the course had been an optional course in Theory of Medical Science run by Stig Andur Pedersen, a philosopher, Raben Rosenberg, a psychiatrist and Henrik R. Wulff, a gastroenterologist and this group had been active behind the introduction of the course in the curriculum at the medial school. But now the course had become obligatory with exams and I was called upon by the dean to start it up. The course was placed on the 4th semester and comprised of general philosophy of science, e.g. the problem of demarcation between science and pseudo-science, specific philosophical problem in medicine, e.g. concepts of health and disease, and medical ethics. The textbook for the course was Wulff H. R., Pedersen S. A. and Rosenberg R.: Philosophy of Medicine. An introduction (5). I am happy to tell that the course has become a success among medical students and in 1995 I was awarded the Harald Prize (Harald after Harald Bohr, the brother of Niels Bohr) as the lecturer of the year. Among the reasons for the award were: "Peter Rossel has performed a great achievement in building up the subject Theory of Medical Science and to ensure that it is solidly placed in the minds of the students."

By that time philosophy of medicine had been institutionalized as "The Unit of Medical Philosophy and Clinical Theory" at the Medical Faculty, University of Copenhagen.

3. What does your work reveal about Philosophy of Medicine that other academics, citizens fail to appreciate?

It has been a mark of my (and the Units) teaching and research activities to engage in and grabble with real aspects and problems in medicine – mostly, but not exclusively, the ethical aspects of everyday clinical practice and research. And it has been a challenging feature of these activities that one had, so to say, *to leave the armchair* and get in contact with daily clinical practice and research. In writing this I recall Stephen Toulmin's paper "How Medicine saved the Life of Ethics."(6)

My own work has mainly, but again not exclusively, focused on informed consent in therapy and research both in form of moral philosophical as well as empirical studies and at our Unit we have built up a rather strong tradition for doing empirical ethics. My own research within this field of empirical ethics has taken two different paths. One path is historical and contemporary historical case-studies informed by moral philosophical distinctions and concepts of key episodes within medical science, as exemplified by my analysis of The Doctor's Trial. Another case-study I am currently working on will be presented below as an example of what other academics fail to appreciate. The other path is empirical studies in the form of questionnaires and interviews of informed consent in therapy and research. What is distinctive of these studies is that the construction of the questionnaires and the interpretation of data are undertaken within a theoretical ethical framework. This direction of research thus follows the path of my above-mentioned study of the ethical aspects of the decision making process in general practice. A later study done in collaboration with colleagues at our Unit of Medical Philosophy and Clinical Theory is also about informed consent to therapeutic measures in the daily clinical setting. It showed that the Hippocratic tradition and traditional paternalism is still prevalent among general practitioners (7).

Another study in this line of research I would like to mention is an empirical study of patient's perceptions of the informed consent process of a Danish randomized trial, the DANAMI-2 trial, done by my former Ph.D.-student Anne Gammelgaard (8,9). This trial involved patients who suffered an acute myocardial infarction (AMI) and who were admitted to an emergency department in the acute phase of the disease. It was the aim, by means of questionnaires and in-depth interviews, to analyze how the patients experienced the informed consent procedure. Further on this basis to analyze the ethical issues of clinical trials involving AMI patients and to provide guidelines for the consent process of future trial involving AMI patients.

A study in the first path of my empirical research that I currently undertake is a case-study of "Ethical Aspects of Pre-randomization in Clinical Trials. A Case-study: Mastectomy vs. Tumorectomy." With respect to "therapeutic research" the Declaration of Helsinki (1975) contains the so-called "escape clause" – to repeat: "If the doctor considers it essential not to obtain informed consent, the specific reasons for this proposal should be stated in the experimental protocol for transmission to the independent committee." Remember too, that this clause pertains to patients who are, in the legal sense, competent persons. What kinds of reasons were thought to justify the suspending of informed consent for these patients? On this question the final version of DoH (1975) is silent and the "escape clause" thus confers a broad and unspecific authority to the research ethical committees (RECs).

As mentioned, DoH (1975) has been described as the "cornerstone" document pertaining to medical research and as "the most widely recognized source of ethical guidance for biomedical research" (3). As such the document has been the subject of extensive comments and analysis but this most controversial paragraph in DoH (1975) seems to have gone almost unnoticed in the literature or to the extent it has been commented upon, it has been mingled together with the so called "loophole" in DoH (1964) and interpreted as an expression of "the therapeutic privilege" and traditional medical paternalism. But does it *make sense at* all to talk about paternalism when a patient/research subject is enrolled in a RCT without informed consent in accordance with the "escape clause"? This is not just a terminological issue because, if the reason for suspending informed consent in such a context is not a paternalistic reason, the suspension is not done (primarily) out of a concern for the best interests of the individual patient/research subject, but must be an expression of other concerns. Again this is not an idle question since, as mentioned, the Basic Principles in DoH (1975) states: "Concern for the interests of the subject must always prevail over the interests of science and society"(I.5). (For a detailed analysis – see my forthcoming (10))

What kinds of reasons, then, would justify the suspension of informed consent within the framework of DoH (1975)? One important way (and perhaps the only way) to answer this question is to analyze specific research projects carried out in accordance with the escape clause and approved by a REC. One such project was a RCT running from 1982 to 1989 where The Danish Breast Cancer Cooperation Group (DBCG) undertook a study comparing mastectomy vs. tumorectomy in case of breast cancer (Protocol DBCG 82-TM). The study was designed according to a Zelen-design with pre-randomization and one-armed information to the experimental group (tumorectomy). None of the patients/research subjects were, however, informed about the randomization, thus suspending the requirement of informed consent. The RCT was approved by The Central Scientific Ethical Committee of Denmark (CVK) in 1982. The DBCG 82-TM thus exemplify the reasons the researchers and the CVK considered sufficient to suspend informed consent. (This project is not the only one where the CVK approved a Zelen-design suspending the requirement of informed consent by not informing the patients/research subjects about their participation in a randomized study).

By examining the reasoning of the CVK in approving the DBCG

82-TM protocol, one have to conclude that the overriding reason was "that the trial satisfies the scientific requirements to such a study, which again is an absolutely necessary precondition for that the results can be utilized to the benefit of the patients." (For a detailed analysis – see my forthcoming (11)). I also have to conclude that the concern I raised in my book Medicinsk Etik (2) that the "escape clause" might legitimate that the interests of science and society could prevail over the concern for the interests of the individual patient/subject within the framework of DoH (1975) thus had been vindicated.

The study was carried out according to a Zelen-design and it is of special interest to look at the problem that pre-randomization (or randomized consent as it is also called) is an attempt to solve. According to Zelen "One of the principal reasons why clinical investigators decline to participate in randomized studies(is) that they believe that the "patient-physician relation" is compromised in complying with federal regulations." (12).The reference is to Title 45, Code of Federal Regulations, part 46 (U.S.A. regulation) where the requirement of informed consent is formulated in relation to RCT's. This requirement imply that the patients have to be informed about randomization and because of this "the physician-patient relationship could be compromised if the physician makes it known to the patient his/her inability to select a preferred therapy." (12)

How could the patient-physician relation be compromised by information about randomization? A constituent element *in the relationship of trust* in the ordinary patient-physician relationship is the patient's justified expectation that the physician – in accordance with the Hippocratic tradition – only has the patient's best interests in mind in recommending a therapeutic measure. When involving a patient/research subject in a RCT, however, this cannot possible be the case. If the RCT has to be ethically justified at all, the physician/researcher ought not to know (equipoise) which method of treatment is in the patient's best interest. Furthermore, randomization is a research component that is foreign to the ordinary patient-physician relationship. How, then, is the Zelen-design thought to solve the problem of not compromising the patient-physician relation? Zelen's answer is as follows: "The proposed new design has the desirable feature that the physician need only approach the patient to discuss a single therapy. The physician need not leave himself open, **in the eyes of the patient**,(my emphasizing)to not knowing what he is doing and "tossing a coin" to decide the treatment. Thus the patient-physician relation is not compromised. On the patient's side, there is also an important advantage: before providing consent the patient knows which treatment will be given." (12)

From a systematic moral point of view I find this extremely interesting. In order not to endanger the relationship of trust encompassing the patient's expectation that the physician only has his/her best interests in mind in his advice about treatment, the physician/researcher withhold the information that he does not know which treatment is in the patient's best interest and, furthermore, withhold information about randomization as deciding the choice of treatment. Such a strategy of deception can hardly be universalized.

Also from the view of public policy such a strategy is interesting. Can it survive publicity? If it becomes known in the public (as it did) would it not endanger the relation of trust in general between the medical profession and the public at large?

Actually, when the study reached the public in 1989 there was an extremely heated debate about randomized trials and the study was one main factor in a process that lead to legislation in this area in Denmark. A law without the above mentioned escape clause. And this is another reason why I find the case-study of DBCG 82 -TM fascinating: Just another controversy in the history of human subject research that brought about a development in the ethical regulation of research on human beings. (Again, for a detailed analysis – see my forthcoming (12))

Now, finally, to my answer of the question about what my work reveal that other academics/citizens fail to appreciate. I find that academic philosophers mostly do not appreciate the empirical approach to ethics outlined here. It is not philosophical enough, so to say. On the other hand, I find that historians and social scientists mostly do not appreciate the approach of analyzing historical and contemporary historical episodes within a framework of moral philosophical distinctions and concepts. I do find, however, that doctors in general do appreciate this empirical cum philosophical approach.

Why, then, do I find this empirical cum philosophical approach a fruitful one – apart from being interesting per se? The answer is that the empirical research can make one reflect more thoroughly on the moral philosophical justification of informed consent, both in daily clinical practice and with respect to human subject research. One of my findings in the study of the ethical aspects of the decision making process in general practice was that a reason among general practitioners for not acting paternalistically, and thus without the patient's consent, was not respect for some kind of *individual autonomy*, but because the paternalistic act in question would violate *an agent-relative deontological constraint* against lying for or deceiving the patient. And with respect to human subject research it strikes me that the main reasons why the research on human beings in the concentration camps is moral objectionable was not because of the violation of some kind of *individual autonomy* but, apart from, of course, the cruelty of these often invalidating and deadly experiments, that they involved *coercion* of involuntary research subjects. And take the notorious Tuskegee-study. Again, what strikes one as morally objectionable is not a lack of respect for some kind of individual autonomy, but that the subjects were *deceived* into believing, that the were offered *treatment* for syphilis – and objectionable, of course, because these research subjects were denied treatment, when penicillin later had been proved an effective treatment. Finally, there is my own case-study of the DBCG TM-82. Again, what is problematic from a moral point of view is not lack of respect for some kind of individual autonomy, but that the women were *deceived*. Again, of course, there are essential morally relevant differences between this study and the Tuskegee-study. In the DBCG TM-82 study the women in the control group received the standard therapy, i.e. removal of the whole breast, and the experimental group was offered breast preserving surgery with the removal of only the tumor lump. Information about that they were enrolled in a randomized study was, however, explicitly withhold from the patients/subjects.

To sum up: The reasons why the just mentioned kinds of research on human beings without their informed consent are objectionable or problematic is that the research subjects were either coerced or deceived. Interestingly these findings accord well and lend support to Onora O'Neill's more philosophical argumentation for the justification of informed consent, to quote: "Rejecting and avoiding coercion and deception are of ubiquitous and fundamental importance in ethics, and especially in bioethics. One advantage of taking them seriously is that, taken together, they provide the basis for informed requirements: Action that either coerces or deceives others stand in the way of free and informed consent; conversely where free and informed consent is given, agents will have a measure of protection against coercion and deception". (13)

4. What, if any, practical and/or social-political obligations follow from studying medicine from a philosophical point of view?

Primarily I think that I have an obligation in my lectures for medical students to enable them to identify and analyze the ethical aspects of daily clinical practice and research – issues the will have to deal with later as physicians. Furthermore, to give them an impression of the complexity of some of these issues and most importantly to enable them to carry through a consistent and coherent reasoning in order for them to arrive at a rationally justified position on some of these issues. In order to do this in a qualified way, I myself is under the obligation to be in touch with and have a familiarity with clinical realities outside the purely academic world. Actually, this is a rather delightful obligation. I find the cross-fertilization between the real problems in the real world of medicine and moral philosophy the most satisfying feature of working within the field of medical ethics.

I also find that I have an obligation to communicate my research to a wider audience, especially within the health care system. And during the years I have given lectures and been in discussion with nearly all sections and professions in the health care system – likewise a delightful obligation.

Finally I think one has the socio-political obligation from time to time to intervene in the public debate, e.g. when a new bill is introduced in the parliament or existing legislation proposed amended. The aim of the intervention should be, I find, to try to eliminate or reduce gut reactions and the YUK-factor and to pinpoint inconsistent and incoherent proposals and arguments.

5. With respect to present and future inquiry, how can the most important problems concerning Philosophy of Medicine be identified and explored

With respect to future inquiry a fruitful approach might be to extend the experimental approach in philosophy to philosophy of medicine and medical ethics. Experimental philosophy or for short X-phi is a recent flourishing movement in analytic philosophy. Its logo is a burning armchair. Instead of just consulting your own intuitions in the armchair about classical philosophical concepts and problems as for example determinism, free will and moral responsibility as well as moral dilemmas, philosophers also ought to explore in a systematic way so-called "folk concepts" and "folk intuitions" about contested philosophical problems. Within X-phi one wants to replace or supplement the armchair as a method with methods from e.g. experimental psychology, social psychology and neurobiology (e.g. fMRI) in the study of folk concepts and folk intuitions concerning classical philosophical questions and moral dilemmas.

An illuminating example comes from action theory and the relationship between intentional action and moral judgments. From a moral point of view it is ordinarily assumed a crucial question in the assessment of an act whether or not the action was intentional. And the perhaps most striking experiment in X-phi is a case study of folk concepts of intentional action – an experiment with surprising results. In the experiment, which has been repeated several times, the American philosopher Joshua Knobe (14) presented "folk" with the following scenario that can be named *"harm"*:

> The vice-president of a company went to the chairman of the board and said, "We are thinking of starting a new program. It will help us increase profits, and it will also harm the environment."
>
> The chairman of the board answered, "I don't care at all about harming the environment. I just want to make as much profit as I can. Let's start the new program."
>
> They started the new program. Sure enough, the environment was harmed.
>
> *Did the chairman harm the environment intentionally?*

Knobe discovered that 82% of the subjects responded that the chairman intentionally harmed the environment. Other subjects were given a corresponding scenario but with one difference that can be named *"help"*:

> The vice-president of a company went to the chairman of the board and said, "We are thinking of starting a new program. It will help us increase profits, and it will also help the environment."
>
> The chairman of the board answered, "I don't care at all about helping the environment. I just want to make as much profit as I can. Let's start the new program."
>
> They started the new program. Sure enough, the environment was helped.

Did the chairman help the environment intentionally?

Kobe found that only 22% of the subjects responded that the chairman intentionally helped the environment.

This striking *asymmetry* is known as the Knobe Effect. It is a robust but paradoxical finding. Knobe's own interpretation of the effect is that folk intuitions about whether an action is intentional or not is influenced by or depend upon whether the outcome of the action is good or bad. In other words that folk intuitions and folk concepts about intentional action depend upon the moral assessment of the action and its side effects.

For, at least philosophers, Knobe's finding is surprising because most philosophers have the conception that moral judgments can't determine whether or not an action has to be classified as intentional. The traditional conception of the relationship between intentional action and moral judgments is, that first one tries to establish whether an action is intentional and then makes the moral judgment. Knobe's finding, however, suggests that the causal arrow also goes in the other direction: From moral judgement to intentional action. Here, then, we have a case where X-phi has found a conflict between philosophers intuitions and concepts and folk intuitions and concepts.

The notion of intentional action plays a central role in the analysis and discussions of issues in medical ethics. For example is the distinction between acting intentionally and unintentionally seen as crucial, conspicuously in the euthanasia debate. In the administration of pain relieving medication and/or sedatives with the possibility of hastening death it is ordinarily seen as crucial for the moral (and legal) assessment whether the hastening of death was a merely foreseen, but unintentional side effect or the doctor intended such acceleration. The former is under certain circumstances seen as permissible whereas the latter is forbidden. According to the usual conception the causal arrow has one direction: From intentionality to moral judgment and not the other way around. The Knobe Effect, then, suggests experimental medical ethical studies about whether or not the causal arrow also go in the other direction concerning end of life decisions: From moral judgment to intentional action.

In addition to the concept of intentional action, Knobe has the understanding that his effect reflects a general phenomenon that does not just bear on the concept of intentional action. If this hypothesis is correct, one has to expect that other important concepts likewise are influenced by or depend upon moral judgments, and if this is the case his hypothesis seems relevant and interesting for other distinctions in medical ethics. One such distinction is the distinction between *doing and allowing judgments*. An example of this distinction is the distinction between killing and letting die where this conceptual distinction likewise has been seen as morally relevant in questions about active and passive euthanasia. The traditional view is that it can be morally permissible to allow a patient to die (passive euthanasia) but always morally impermissible to kill the patient in spite of well-considered wish to die (active euthanasia). In other words, if one actively terminates the patients life this is morally forbidden. On the other hand, if one let the patient die e.g. by withholding treatment it is under certain circumstances morally acceptable. In the traditional view the causal arrow thus goes in one direction: From whether the way of acting was a doing or an allowing to the moral judgment and not the other way around. Analogously with the concept of intentional action, it would be interesting to do experimental work in medical ethics to test whether the arrow also goes in the other direction.

Within experimental philosophy conflicts between philosopher's intuition and "folk intuitions" have been uncovered with respect to intentional action and might also be uncovered with respect to other distinctions of relevance for issues in medical ethics. It might therefore be relevant and interesting to expand the domain of experimental philosophy to medical ethics. With respect to end-of-life decisions, it will be interesting to undertake experimental studies of "folk intuitions" about intentional action and the distinction between doing and allowing. "Folk intuitions" should include lay persons' intuitions as well as persons working within the health care system, who have to deal with these issues in their daily clinical work. It might be interesting, furthermore, to explore whether and to what extend it is the use of *moral heuristics* that generate "folk intuitions". Moral heuristics are a sort of rules-of-thumb that are doing well most of the time but by overgeneralization might misfire (15). Moral heuristics might be seen in the light of *dual-process theories* (16) that distinguish between cognitive operations that are rapid and associative from others that are slow and reflective, and the growing evidence in moral psychology for people often make automatic and unreflective moral judgments (17).

Such studies might contribute to a better understanding of contentious issues in medical ethics.

It might also be interesting to explore how the experimental approach in philosophy can be extended to other topics of specific interests in philosophy of medicine, e.g. personal responsibility for health, concepts of health and disease, and causation in medicine.

References.

1. Rossel, P.: Lysenko-affæren I: Nørreklit, L. (red.): Videnskabshistorie, Aalborg Universitetscenter, 1977: 89-125.

2. Rossel, P.: Medicinsk Etik. København: G.E.C. Gads Forlag, 1979.

3. Robert V. Carlson, Kenneth M. Boyd & David J. Webb: The revision of the Declaration of Helsinki: past, present and future. Br J Clin Pharmacol 57:6, 695-713. 2004

4. Rossel, P.: Autonomi, paternalisme og den typiske danske læge I: Holm, S. & Lützen, K. (red.): Ekspert og medmenneske, København: Akademisk Forlag, 1997: 23-39.

5. Wulff, H.R., Andur Pedersen, S., Rosenberg, R.: Philosophy of Medicine. An Introduction. Oxford: Blackwell Scientific Publications, 1986.

6. Toulmin, S.: How Medicine saved the Life of Ethics. Perspective in Biology and Medicine, 24 (1982):736-750.

7. Krag, A., Nielsen, H. S., Norup, M. Madsen, S.M., Rossel, P.: Research report: do general practitioners tell their patients about side effects to common treatments? Social Science and Medicine, 59 (2004), 1677-1683.

8. Gammelgaard, A., Mortensen, O. S., Rossel, P.: Patients' perceptions of the informed consent in acute myocardial reserach: a questionnaire based survey of the informed consent process in the DANAMI-2 trial. Heart, 2004; 90: 1124-1128.

9. Gammelgaard, A., Rossel, P., Mortensen, O. S.: Patients' perceptions of informed consent in acute myocardial infarction research: a Danish study. Social Science and Medicine, 58 (2004), 2313-2314.

10. Rossel, P.: The "escape clause" in the Declaration of Helsinki 2, 1975: An expression of traditional medical paternalism? (forthcoming)

11. Rossel, P.: Ethical aspects of pre-randomization in clinical trial. A case-study: Mastectomy vs. tumorectomy. (forthcoming)

12. Zelen, M.: A new design for randomized clinical trials. N Eng J Med, 1979; 30:1242-1245.

13. O'Neill, O.: Autonomy and Trust in Bioethics.Cambridge: Cambridge University Press, 2002.

14. Knobe, J: The Concept of Intentional Action: A Case Study in the Uses of Folk Psychology. *Philosophical Studies*, 2006; 130:203-231. Reprinted in Knobe, J. & Nichols, S. (eds.): Experimental Philosophy. Oxford University Press, 2008.

15. Sunstein, Cass R.: Moral Heuristics. Behavioral and Brain Sciences. Cambridge University Press, 2005.

16. Evans, Jonathan, St.B.T.: Two minds: dual-process accounts of reasoning. Trends in Cognitive Science. Vol.7, No.10, October 2003.

17. Haidt, J.: The Emotional Dog and its Rational Tail: A Social Intuitionist Approach to Moral Judgment. Psychological Review, 2001; 109(4):814-34.

15

Udo Schuklenk

Professor of Philosophy and Ontario Research Chair in Bioethics
Queen's University, Kingston, Canada

**1. Why were you initially drawn to Philosophy of Medicine? and
2. What does your work reveal about Philosophy of Medicine that other academics, citizens typically fail to appreciate?**

My initial interest in medical ethics and bioethics was both accidental and not accidental. I need to go back here all the way to my philosophy undergraduate years at Bochum's Ruhr-University in Germany. At the time my coming out as a gay man coincided with the reported beginning of the HIV/AIDS pandemic. Gay men were among the groups that were initially most significantly affected by HIV/AIDS in the Western hemisphere. Conspiracy theories abounded in the gay community, while medical ethicists and theologians were busily discussing 'ethical issues' such as whether AIDS was 'God's punishment for an unnatural lifestyle', whether or not we should quarantine people with AIDS and other questions that I considered mostly irrelevant. From a philosophy of science perspective I got drawn into a debate that was raging at the time, namely around the question of whether or not it had been proven that HIV is the sole cause of AIDS. At the time a reasonably clear correlation existed, but correlation must not be conflated with causation, as those sceptical of what was at the time the HIV-AIDS hypothesis were busy pointing out. Much later I published a paper on the professional obligations biomedical scientists - who subscribe to the view that HIV is not the cause of AIDS - with regard to a scientifically untrained audience. Were their attempts at converting a lay audience to their controversial views ethically acceptable, considering that government policies based on their views led to more than a 100,000 lives lost to AIDS unnecessarily in South Africa?

Prior to my intellectual engagement with AIDS I had devoured Peter Singer's *Animal Liberation* as well as his *Practical Ethics*. I admired the inescapable logic of the argument presented in *Animal Liberation* as well as Singer's writing style. This was the opposite of what German philosophy was all about. It was unpretentious and it had questions at its heart that were relevant to our daily lives. It is fair to say that without reading those books at the time, it is unlikely that I would have continued with my philosophy undergraduate studies. I was disappointed by the content of what was on offer during my undergraduate years, and would likely have moved on to something else. As an undergraduate student I was interested in big philosophical questions and themes that mattered practically to how we live our lives. I wondered whether we had moral obligations to future generations, whether non-human animals had moral standing; as an elected Green politician I cared about environmental ethics including the morality of burdening future generations with the fall-out of our reliance on nuclear power. In my university most of the professors lectured us for months on end about their research on something or other that Leibniz or Hegel may or may not have said. They rarely ever had more than 10 students in lecture theatres designed to hold hundreds of students. I could not help but wonder whether I was witnessing the final days of a dying discipline. I appreciated the types of questions they were invested in but it was also evident to me that, whatever the answers to the questions they were investigating at great expense, it would make no difference to anyone's life.

At the same time there was a big controversy in Germany about Peter Singer's views on infanticide and 'lives not worth living'. It did not help his case in Germany that the translator of his work chose a phrase similar to one the Nazis used in their propaganda during their mass murder of disabled people. The international reaction, both negative as well as very supportive of the Australian philosopher, made me wonder whether there was hope for philosophy after all. Here was someone who clearly thought about the types of philosophical questions I cared passionately about, and the wider public equally passionately discussed what he had to say. My political friends in Germany were horrified that I refused to call Singer's name and demonstrate with them against his views. I recall well, how I ended up in a meeting of particularly paranoid Singer foes where someone checked my Walkman (yes, Walkman players were en vogue in those days) to ascertain whether I was secretly taping their scheming against a Singer lecture at a German university. Meanwhile every single one of my own philosophy professors failed to stand up for his academic freedom right to speak his mind on German university campuses. Philosophy in Germany proved to be a dead-end for me for another reason. In order to complete a graduate philosophy degree in Bochum I would have needed at the time formal Latin qualifications. I was not prepared to study Latin for the sake of meeting the pointless requirements of that university's philosophy department.

Not sure what to do next I headed to Australia for a lengthy backpacking tour across that great country. During a particularly bad spell of weather in Melbourne I decided to head out to Monash University to visit Helga Kuhse and Peter Singer. I had a bone to pick with Peter Singer about arguments in *Practical Ethics* that I disagreed with him on. Not realising at all at the time how famous Peter Singer was I took as a given that he would want to see me and discuss my views of his work with me. I never suffered a lack of self-confidence it seems, even in those days as a failed German philosophy undergraduate student. In his usual unassuming way Singer agreed. We had an hour or two worth of cordial conversation; I conceded defeat; and eventually he asked in passing, over lunch, whether I would be interested in doing a Ph.D. at Monash University. I liked Australia a great deal, so upon my return home I worked toward securing grant funding both in Germany and in Australia to undertake a research PhD under his supervision. No doubt, our chance encounter changed the course of my life for good. He prudently discouraged me from doing work on my proposed thesis topic, assisted dying, arguing that more or less the relevant arguments on both sides of the divide had been made and that there was little that I could contribute to this debate in a doctoral bioethics research thesis that would be truly original. This advice, both farsighted and correct, was accompanied by the suggestion that I should consider particular ethical issues related to the growing AIDS pandemic. That is what I did. I still pass this kind of advice on to my own students today: Do not try to engage – especially as a new entrant to the field – a topic that has exercised more experienced and possibly greater minds than yours for a long time and assume that you would likely have something innovative to contribute. There might be the odd exception to this general rule, but as general strategy advice, it makes sense.

My doctoral thesis was not motivated by a purely theoretical interest. I subscribe to the view that in applied ethics our research interests are more often than not motivated by our personal experiences, convictions and feelings as opposed to abstract concerns. It is possible but unlikely that you would spend years of your life undertaking research on world poverty if you did not actually care personally and passionately about the detrimental effect poverty has on people's quality of life and on their equality of opportunity. Students researching theses on animal rights or animal welfare are more likely than not personally as opposed to just academically interested in the subject matter. Unless you are concerned about animal suffering you are not likely to spend years of your life researching a thesis on the moral status of animals. Feminist philosophers have long recognised the importance of one's personal interests and experiences for and in one's research, while mainstream analytical philosophy tends to pretend – against the available evidence – that it is pure reason and logic alone that should drive a good philosopher's inquiry. To my mind personal investment and sound method can and should go hand in hand. My personal interest drove my research agenda. Among AIDS activists at the time the clinical trials' system was a great source of concern. Actually, that is an understatement, people with AIDS protested aggressively against this system. In a nutshell, the contested trial method required dying AIDS patients to accept enrolment in a placebo controlled clinical trial or go without access to promising experimental drugs. The scientific rationale for this method was sound, the ethics of it was far from self-evident. No gold standard of AIDS clinical care existed at the time, hence the research objective was to find out whether a particular experimental agent was more or less likely to improve patient survival and/or quality of life, as opposed to providing nothing. That was the theory. In practice, desperate, well-educated, usually young AIDS patients fighting for their lives understood what the randomisation in the placebo arm would mean for them individually: death. Many of them were prepared to accept the risks involved in trying the experimental agent, simply because they knew from their dying friends around them that the placebo would not assist their survival.

Here we had a conflict between sound clinical research methodology, and the demands of dying patients afflicted by a catastrophic illness, to access experimental drugs. My thesis came down firmly on the side of the patients' demands, much as I appreciated the methodological argument in favour of placebo controls. The view argued for in the thesis was one that I still hold today: We must not hold desperate dying patients for ransom in order to undertake placebo controlled clinical trials. Society needs to ensure that those who enrol in such trials are true volunteers. This requires a reasonable alternative to the possibility of getting randomised into the placebo arm when one does not want to be in that arm. It is fair to say that my views were shared at the time by very few bioethicists. Most straight bioethicists, by and large personally disconnected from the fight for survival that AIDS patients were engaged in, supported the clinical research establishment. They lacked a clear personal understanding of the situation these patients were in. Importantly, they argued their case in spite of strong empirical evidence that AIDS patients cheated in such significant numbers in these clinical trials (by means of sharing medication, for instance) as to render them useless. It did not surprise me that someone who is in the fight for his very survival will not let himself be bound by the niceties of informed consent forms. He is likely to promise anything and everything to join the trial with the experimental drug he thinks is most likely to assist his quest for survival. I rejected the mainstream bioethical analysis that suggested that these patients were giving voluntary first person informed consent hence they were bound by the deal they signed on to when they signed the trial consent forms. The voluntariness premise of this argument was implausible. These patients were anything but volunteers, given that access to the very same experimental agents tested in these trials was unavailable to them outside the placebo-controlled trial. No doubt, without my personal interest and personal involvement in AIDS activism I would have written a very different thesis to the one I wrote at the time. Peter Singer patiently supervised my thesis writing and managed to calm my often-angry prose down so that a successful doctoral research thesis was eventually produced within the funding period my grants provided for. It is probably fair to say that unlike any thesis I could have written on assisted dying, my thesis – even if you disagree with it - constituted a thoroughly original contribution to the ongoing debate about the ethics of placebo controlled trials in catastrophic illness. Singer's advice on my dissertation topic proved to be spot-on. I also discovered that death does indeed focus the mind on things that matter.

Despite taking out Australian citizenship I decided that 'retiring' in one academic institution, even one in a city as liveable as Melbourne, was not what I wanted to do. I also hoped that as an academic researcher and teacher I could be of greater use in a developing country than I could ever be as a bioethicist in Australia. I applied for the headship of a to-be-established bioethics division at the Witwatersrand University Faculty of Health Sciences in Johannesburg. Under the leadership of Max Price (at the time of writing the Vice Chancellor of the University of Cape Town) the faculty was determined to improve its bioethics curriculum and thankfully it was determined to attract me to the job. I took charge of establishing the new division as well as creating from scratch curricula for medical and dental students as well as students of the allied medical disciplines, eventually, ably assisted by my growing staff complement, I put in place the first general bioethics Master's level graduate degree program offered in an African medical school. Moving to South Africa gave my intellectual growing-up a warp-speed type boost. I realised again very personally how the confrontation with death truly focuses the mind. For the first time I saw with my own eyes large numbers of people with AIDS dying preventable deaths because of government inaction as well as exorbitant drug prices. While most of my bioethics colleagues in developed countries were concerned about human cloning, transhumanism, neuroethics and any number of ethical problems predicated on scientific developments that may or may not eventuate, I focused on ethical questions about drug prices, the morality of the international patent regime, resource allocation issues and public health ethics. I held and still hold the view that the number of people significantly affected by a particular issue matters with regard to the choice of ethical problems we ought to focus on intellectually. Contrary to many of my colleagues I do not accept the argument that the problems in these – usually hi-tech, futuristic areas are theoretically so significantly more challenging and valuable than the theoretical problems we encounter for instance in public health ethics that we should continue researching them. Consequently I became ever more disillusioned with my field's preoccupation with what I hold to be largely inconsequential questions.

Research funders over the last decade or so have poured unprecedented amounts of money into research ethics. Much of this is ostensibly directed at ethical issues in clinical research in developing countries. Western bioethicists, like other academics under ever increasing pressure to generate income, have been travelling the developing world ever since, 'training' and lecturing people there on what they consider to be appropriate standards of care in clinical research and such matters. Having lived for five years in a – reasonably wealthy – developing country, I am concerned about the ethics of wealthy academics from developed countries (with access to superb health care), who stay in upmarket hotels in developing countries, busily writing papers in which they create ever newer reasons for limiting developing world trial participants' access to medicines. Oftentimes these papers are peppered with stories about the locals, no doubt to add authenticity. In some ways this behaviour mirrors the behaviour of said academics who supported the placebo trial system in the West. There was a personal disconnect from the patients whose lives were at stake. The same is true to a large extent of our world travelling Western research ethicists. I suspect that it might be true that if some of these colleagues jet-setted less around the world and instead spent more time living in such countries their academic outputs would look significantly different. Unsurprisingly perhaps, most but by no means all bioethicists hailing from developing countries – while frequently struggling professionally with the deficiencies of a less than ideal tertiary education and lack of access to relevant literature – see themselves as proud advocates on behalf of impoverished patients. There is no pretence of impartiality on their part. It is remarkable that for many of the impartial Western research ethicists their impartiality leads them straight to uncritical support of the policies of their funders, such as for instance the US NIH. The personal experience of seeing impoverished patients die preventable deaths focused my mind in ways that would not have happened had I stayed in Melbourne. Ever since I have tried to address ethical questions that have a significant policy import and that affect significant numbers of real people in lieu of questions that are theoretically interesting but ultimately inconsequential. My view, unlike that of many of my colleagues, is that if you 'do' applied ethics, make sure you apply your critical thinking to real-world problems, not problems cooked-up in her study by an armchair philosopher in search of a problem to test her analytical tools.

One final thought on the personal involvement matter. It would be understandable if what I have said up to this point would be misunderstood as suggesting that ethical analysis is merely a means of supporting already arrived at conclusions. That is not the case. To give you just one example: I recently co-authored a paper on global health obligations. Prior to working with Christopher Lowry, then one of our PhD students, on the manuscript I took as a given that need should be the crucial criterion for the allocation of scarce resources. When our manuscript was completed I had changed my mind on this issue. Our investigation discovered that if we want to be serious about treating equal interests equally, and about maximising positive health outcomes in the context of just resource allocation decision-making, ease of prevention should trump individual need in most cases when it comes to global health aid.

3. What, if any, practical and/or social-political obligations follow from studying medicine from a philosophical point of view?

I take this question to ask whether the insights gained from undertaking research should be tied to a kind of professional obligations on part of academics to act in order to influence policy on the matter investigated. I do think that well-meaning, educated people can reasonably hold different views on this topic. Some hold the view that academics should investigate freely any particular matter within their areas of specialist expertise, and then publish whatever their research outcome is. Respect for academic freedom demands this much. The publication should take place ideally in one academic forum or another, but not in mainstream media, the internet or other fora. The academic in this scenario does quite deliberately not leave the academic ivory tower. In the humanities this baseline view often leads to two different stances: 1) The view that we should in our academic outputs provide balanced reviews and refrain from taking a strong stance for or against any particular substantive view. Basically we would aim for balance as opposed to controversy. 2) The obvious alternative view is that it is acceptable to take a strong stance. Common to both views is the assumption that academics should refrain from going public. It is acknowledged, usually with some trepidation, that there is a chance that others might use their work for their own policy objectives, but academics themselves should not get personally involved in the grubby world of political campaigning. The idea of academics as public intellectuals is very strongly frowned upon by proponents of this set of views about the academy. These colleagues believe that the academy's standing is undermined if the wider public came to the conclusion that academics are partisan in their work, rather than that academics are professional innocent bystanders of a kind who contribute towards clarifying the matters at stake and who provide sound analysis, untainted by ideological interests of some kind or other.

I do not subscribe to this view, because I hold it to be impossible for academics in our field to undertake research without having some kind of what Jurgen Habermas calls sensibly our 'Erkenntnis leitendes Interesse'. I much prefer academics taking a strong stance, arguing their case vigorously and openly, making clear for all to see what their ideological convictions are. With that being in the open, as opposed to frequently hidden under something like, for instance, the heart warming but vacuous rhetoric of 'human dignity', society can derive a greater benefit from seeing where particular ideological commitments take it with regard to controversial issues in medical practice and health care policy. Accordingly I think that academics, certainly in bioethics, could do worse than sharing their views with wider audiences in newspaper OpEd's, on their blogs as well as on radio and television. I suspect that if academics have any obligations at all, it is to ensure that their research is shared with as wide an audience as possible with a view toward maximising its societal impact. This certainly seems to hold true for academics working in societies where universities are publicly funded.

Generally speaking, I think that academics (not as academics but as moral agents) have the moral responsibility to improve societal well-being. This obligation exists both within the realms of our personal as well as professional lives. It is not a unique obligation derived from some kind of professional academic responsibility. Professional codes of conduct often include passages promising that the professional will endeavour to work toward the greater good of society or humankind, but the moral basis of this promise can not realistically be found in the profession itself, it is grounded in a utilitarian commitment to the maximisation of well-being or in alternative moral frameworks. All other things being equal, against this backdrop a well-chosen academic research endeavour would be characterised by having a better chance of improving societal well-being than an ill-chosen academic endeavour. Moral obligations to act do not follow from the philosophical study of medicine as such, however, the philosophical study of medicine might lead to insights that require academics to act in response to their more foundational moral obligations as moral agents, to improve societal well-being. At least those who subscribe to consequentialist moral frameworks have to accept moral obligations that impact on the types of questions they choose to investigate. These consequentialist moral obligations also impact on how academics should go about disseminating the results of their research.

4. What do you see as the most interesting criticism against your own position in philosophy of medicine?

I hold a variety of positions on different topics in our field. What I find most challenging with regard to criticism of my views is twofold: 1) the criticism voiced often accepts my consequentialist line of reasoning, but reaches diametrically opposing conclusions to mine, based on making different empirical assumptions. 2) These challenges are interesting not only because it is often difficult to reach a consensus on those empirical assumptions, but also because they are pointing out a methodological problem with consequentialist ethical analyses. How certain should we have to be of the likely consequences of our analyses prior to airing them in public?

To give you just a few examples: I argued that genetic research on sexual orientation should not be undertaken, because the risk of harm to gay people in homophobic societies is too great. Ethical review committees should refuse to permit such research to go ahead. The criticism of my position held essentially that the risk is not that great, and that – in any case – the price paid for violating academic freedom is so steep that it would to better permit genetic research on sexual orientation to continue. Unsurprisingly, the latter view was held most strongly by liberal US based academics with a keen political interest in discovering a biological cause of homosexuality. A biological cause would have brought them closer to their parochial national policy objective of getting constitutional protections for gay people akin to the protections ethnic minorities enjoy. Truth be told, neither I nor those criticising me had any method for ascertaining how significant the risk of abuse really is, and the critics had furthermore no method that could demonstrate uncontroversially that the value of academic freedom is of greater significance than avoiding the abuse I believed was inevitable if such a genetic cause was ever to be found. All that I had to offer was evidence of past abuse whenever a biological cause was researched, and I had at least one recent paper in a medical journal where someone discussed the question of testing for a 'gay gene' (if there was one) in the absence of a 'cure' for homosexuality. Certainly my evidence constituted anything but watertight proof that there was a significant risk of abuse. My opponents had precisely nothing to support their contention that the abuse risk was negligible. Crucial methodological questions remained: What does the absence both of relevant data as well as the absence of a method aimed at balancing the risk of abuse (or actual abuse taking place) against the value of academic freedom of research tell us about the quality of our analyses? It was water on my mills that, years later, a straight bioethicist proposing that a duty of procreative beneficence exists on part of parents to maximise their off-spring's likelihood of living an optimally happy life. He suggested that if a prenatal test for homosexuality existed in homophobic societies it should be used to reduce or eliminate the number of homosexual off-spring. He simply aimed at improving overall societal well-being and suggested that straight kids in such societies likely would enjoy a better quality of life. He also duly gave the politically correct nod toward reducing societal homophobia, and to be fair, I doubt that homophobic motives drove his agenda. Never spelled out in my straight colleague's analysis was what the consequences for gay people would likely be in a society where prenatal diagnosis for homosexuality became a legitimate instrument of reproductive planning. Arguably the standing of gay people in such a society would be weakened and their quality of life reduced. It is questionable how this bioethicist's nod toward combating homophobia, and his tacit support for enabling homophobic societies to act on their prejudices, can be squared. None of this is acknowledged to be a problem by our straight bioethicist and his impartial analysis. Unsurprisingly, this issue did not even register in his analysis. As so often occurs in bioethics, in this case a whole range of empirical assumptions are being made (the evidence is unavailable), and ethical as well as policy recommendations were happily made in the absence of sound data. Good bioethics must do better than this.

A second example. Critics of my views on placebo controls in trials involving terminally ill patients argued that such trials must go ahead, and that we could not afford diluting results by permitting desperate patients to access experimental drugs by other means. I had strong empirical evidence that desperate patients cheat in such significant numbers in such trials as to render their predictive value questionable if not useless. They thought that they were coerced into trial participation by means of coercive offers. The counter argument held that adopting my ethical stance would result into much longer recruitment processes, resulting in avoidable loss of human life among patients waiting desperately for life-preserving medicines to finally be found. It is fair to say that both sides probably have a point, but it is difficult for ethicists both to do the numbers and also to determine how significant an ethical price is being paid depending on what line of reasoning is being followed. It is also not just about numbers. After all, autonomous dying patients are pressured into participating in placebo-controlled trials by way of preventing access to experimental drugs by alternative means. Ethically speaking, what is worse (or better): The violation of patient autonomy or executing a (sound) scientific strategy (aimed at serving the greater good of society). In this case we neither had the data, nor a clear method to determine which strategy is more ethical than the other. Here the argumentative stand-off meant disagreement on facts as well as disagreement on the importance of particular values.

At least in some cases the critics of my views were proven wrong, simply because their empirical claims could be proven wrong. I argued very early on that – what is today standard operating procedure – people who seroconvert (ie become infected) in HIV vaccine trials ought to be compensated for the worsening of their clinical baseline by means of providing them with guaranteed access to antiretroviral medicines while they derive a clinical benefit from such drugs. My view took into consideration that in all clinical trials we have participants suffering from therapeutic misconceptions in some form or shape. We truly are unable to detect those patients reliably. It stands to reason that for some such patients their HIV infection constitutes a trial-related injury that ought to be subject to compensation. Given our inability to distinguish such trial participants reliably from those who would have become infected due to informed risk-taking, I proposed that everyone sero-converting in HIV prevention trials in developing countries ought to be provided with access to life-preserving HIV medicines. Critics countered that this would make vaccine trials difficult to finance. The delays caused by the search for additional funding would mean a delay by way of bringing an HIV vaccine eventually to the market, hence more lives would be lost to AIDS than would have been if trials could be had cheaper. Well, in this instance it turned out to be the case that it was not difficult at all to find the additional funding for the treatment of participants in HIV prevention trials. There is no evidence that the treatment requirement actually resulted in the predicted delays. Another counter argument conceded the problem I highlighted but insisted that given that we are unable to determine who got infected due to informed risk-taking and who got infected due to a therapeutic misconception, we should simply not treat anyone at all. We owe this insight to a Western ethicist arguing that no life-preserving clinical care to impoverished, sero-converting HIV vaccine trial participants in developing countries is required, at least not based on the rationale I suggested. Let me point out, not just rhetorically, that this ethicist, if he had chosen to participate in such a trial and if he had sero-converted, would have received clinical care as needed from his developed world health care system. He left open the possibility that there might be alternative ways to justify providing clinical care to sero-converting patients in such trials, alas to the best of my knowledge he has not yet offered such an alternative. He and I agreed at least on the facts on the matter. You decide whether my stance or my fellow bioethicist's stance is more reasonable.

I find all these criticisms interesting because their common denominator seems to be disagreement not only about values (that we can tackle in ethics papers) but also often about facts (that we usually cannot address exhaustively in ethics papers). This suggests to me that perhaps we should aim for more interdisciplinary or multidisciplinary work. I will turn to this in the last section of this chapter.

5. With respect to present and future inquiry, how can the most important problems concerning Philosophy of Medicine be identified and explored?

This is a difficult question, because it leaves open another question, namely 'what is it that makes a problem an important problem?' There seem to be at least two possible interpretations, both of which are legitimate. One interpretation suggests that we should choose problems based on the theoretical significance of the matter at hand. Clearly the theoretical significance of a problem is not necessarily connected to its practical significance. This brings me to a second possible interpretation of important: 'Important' is a problem the solution of which would affect large numbers of people or others with the capacity to experience better or worse lives. I think it is perfectly acceptable for a philosopher to opt for the first of these two interpretations. In line with my earlier analysis of what constitutes valuable academic work in applied ethics I would choose the second interpretation of important.

How can these problems, however we decide which problem is more important among the problems vying for our attention, be explored? While I am in danger here of being criticised for flogging a dead horse, I see the biggest challenge on the methodological front. Bioethical inquiry is most challenging, interesting and innovative when it is informed by multiple disciplines. In grant applications, one of the favourite phrases interspersed generously in application narratives as well as the method section, is that of 'interdisciplinarity'. In reality I have yet to come across a research project that is interdisciplinary. Usually academics from different disciplines (say ethicists, medical lawyers, economists, geneticists and invariably sociologists) get together for their big interdisciplinary application. A careful look at many such applications reveals that there is no such thing as interdisciplinarity. Geneticists often add ethicists to their grant applications because they were advised by the funding powers that are that 'an ethics component' would increase their chances of success with their application. In reality they tend not to care too deeply about the ethics content one way or another, they wish to undertake their genetics research project. Accordingly, when representatives of these disciplines cobble together their interdisciplinary projects, really what they cobble together are various discipline-based projects that are at best tenuously connected to each other under the larger project framework. I think it is a crucial intellectual challenge for the discipline to make interdisciplinarity work, methodologically and practically. This requires a critically reflective answer to the question of what it is that constitutes true interdisciplinarity as opposed to mere multidisciplinarity. More fundamentally it forces the field to address the question of what methods or methods qualify as viable bioethical approaches. I do think philosophy of medicine could progress significantly if we could get a serious conceptual handle on what we mean by interdisciplinarity generally and with regard to specific research agendas and projects. If we succeeded in giving meaning to our rhetoric our research likely would be more fruitful than it is today, simply because we would see a critical cross-fertilisation between different academic disciplines.

In the absence of true interdisciplinarity in bioethics we will remain in a situation where different disciplines continue to struggle to establish their hegemony over the field – not least in order to grab a bigger share of research funding and university posts. To give just one example of this: At the time of writing bioethics has become something of a battleground, more so in the UK than elsewhere in the world, between ethicists and sociologists. Sociologists in bioethics frequently survey people's opinions both by means of quantitative as well as qualitative means. The information can – but all too often does not serve even that purpose – be useful for subsequent ethical analysis. The result of these sociologists' activities are survey after survey querying what taxi drivers, hair dressers, Anglicans, communists, as well as never-ending convenience samples of health sciences students think about human cloning, AIDS, and any number of random other issues. Often these surveys serve as a means to get a graduate degree in bioethics, one usually taught by academics without serious ethics training themselves. Sociologists pretend here to 'do' bioethics while doing what they are trained to do, namely essentially survey people about their views. Surveying people on their reflective or unreflective views on bioethical issues does neither analyse nor settle relevant normative questions. This surveying does not make a sociologist's work any more bioethics than surveying gynaecologists on matters gynaecology would make her survey gynaecology. The fact that this controversy continues, of course, is a result of the already mentioned lack of a consensus on what constitutes sound method in bioethics. It would be interesting to tease out what exactly, if anything, this sociological contribution brings to bioethical inquiry.

There are some pretty clear indications of criteria that scholars should keep in mind when they try to determine which issues to tackle, significant theoretical work remains to be done on the methodological frontiers in bioethics.

Further readings

U Schuklenk, E Stein, W Byne and J Kerin. The Ethics of Genetic Research on Sexual Orientation. *Hastings Center Report* 1997; 27(4): 6-13.

U Schuklenk. Access to Experimental Drugs in Terminal Illness: Ethical Issues. Haworth, New York and London 1998 (pbck 2000).

U Schuklenk. Protecting the Vulnerable: Testing Times for Clinical Research Ethics. *Social Science and Medicine*. 2000; 51: 969-977

U Schuklenk, RE Ashcroft. Affordable access to essential medication in developing countries: conflicts between ethical and economic imperatives. *Journal of Medicine and Philosophy* 2002: 27 (2): 179-195.

U Schuklenk. Professional Responsibilities of Biomedical Scientists in Public Discourse. *Journal of medical ethics* 2004; 30: 53-60.

U Schuklenk, A Kleinsmidt. Rethinking Mandatory HIV Testing During Pregnancy in High HIV-prevalence Regions: Ethical and Policy Issues. *American Journal of Public Health* 2007; 97(7): 1179-1183.

U Schuklenk, RE Ashcroft. HIV Vaccine Trials: Reconsidering the Therapeutic Misconception and the Question of What Constitutes Trial Related Injuries. with RE Ashcroft. *Developing World Bioethics* 2007; 7(3): ii-iv.

U Schuklenk, C Lowry. Two Models in Global Health Ethics. *Public Health Ethics* 2009; 2(3); 276-284.

U Schuklenk, Jv Delden, J Downie, S McLean, R Upshur, D Weinstock. End-of-Life Decision-Making in Canada – The Report of the Royal Society of Canada Expert Panel on End-of-Life Decision-Making. *Bioethics* 2011 (in press)

U Schuklenk, S Philpott. AIDS: The Time for Changes in Law and Policy is Now. *International Journal of Law in Context* 2011 (in press).

About the Editors

Jan Kyrre Berg Olsen Friis

Jan Kyrre Berg Olsen Friis holds a MA in Philosophy from University of Oslo and a Ph.D. in Science Studies from Roskilde University. Presently he is Deputy Director of MeST - Centre for Medical Science and Technology Studies, Department of Public Health, University of Copenhagen, Denmark. His philosophical interests include observer variability in medicine, perception, intuition and tacit knowledge, and embodiment. Friis is author and co-editor of several books on philosophy of science and philosophy of technology, among these are *A Companion to the Philosophy of Technology*, Wiley-Blackwell; *New Waves in Philosophy of Technology*, Palgrave Macmillan; *Philosophy of Technology: 5 Questions*, Automatic Press.

Peter Rossel

Peter Rossel is associate professor in Philosophy of Medicine and head of the Unit of Medical Philosophy and Clinical Theory, Institute of Public Health, University of Copenhagen. He has been involved in research in research ethics and informed consent since the late 1970s and wrote the first book in Danish on medical ethics (*Medicinsk Etik*, 1979) from a philosophical perspective. He is co-founder of The Danish Society for Medical Philosophy, Ethics and Methodology and is its present president. Among his latest publications is *Evolutionary Theory. 5 Questions* as a co-editor.

Michael Slot Norup

Michael Norup is associate professor and teaches philosophy of medicine at the Institute of Public Health, University of Copenhagen. He is a physician specialized in general practice and has a

Ph.D. in medical ethics. Norup has co-authored and edited several books in both English and Danish, among these are *Evolutionary Theory 5 questions*; *Videnskabsteori for de Biologiske Fag*; *Lige Muligheder for Alle*; *Social arv, kultur og retfærdighed*; and *Det menneskelige eksperiment, om menneskesyn og moderne bioteknologi*.

www.ingramcontent.com/pod-product-compliance
Lightning Source LLC
Chambersburg PA
CBHW021841220426
43663CB00005B/346